ECLECTIC EDUCATIONAL SERIES

KEY

TO

RAY'S™ NEW HIGHER

ARITHMETIC

AF813879

Originally published by
Van Antwerp, Bragg & Co.

This edition published by

MOTT MEDIA

ISBN 10: 0-088062-064-1
ISBN 13: 978-0-088062-064-2
Printed in the United States of America

PRESENT PUBLISHER'S PREFACE

We are honored and happy to bring to you the classic Arithmetics by Joseph Ray. In the 1800s these popular books sold more than any other arithmetics in America, in fact over 120,000,000 copies. Now with this reprinting, they are once again available for America's students.

Ray's Arithmetics are organized in an orderly manner around the discipline of arithmetic itself. They present principles and follow up each one with examples which include difficult problems to challenge the best students. Students who do not master a concept the first time can return to it later, work the more difficult problems, and master the concept. Thus in these compact volumes is a complete arithmetic course to study in school, to help in preparing for ACT and SAT tests, and to use for reference throughout a lifetime.

In order to capture the spirit of the original Ray's, we have refrained from revising the problems and prices. Only a few words have been changed, as we felt it wise. Thus students will have to rely on their arithmetic ability to solve the problems. Also the charm of a former era lives on in this reprinting. Flour and salt are sold by the barrel, kegs may contain tar, and postage stamps cost 3¢ each. Through this content, students learn social history of the 1800s in a unique, hands-on manner at the same time they are mastering arithmetic.

The series consists of four books ranging from Primary Arithmetic to Higher Arithmetic, as well as answer keys to accompany them. We have added a teacher's guide to help today's busy teachers and parents.

We wish to express our appreciation to the staff of the Special Collections Library at Miami University, Oxford, Ohio, for its cooperation in allowing us to use copies of their original Ray's Arithmetics.

George M. Mott
Founder, Mott Media LLC

PREFACE

In the preparation of the following pages, the chief aim has been to serve the teacher. Accordingly, through a great part of the work we have merely indicated the operations required, and have thus saved the space which would be demanded for the full statements of analysis, useful chiefly in recitation drill.

The help afforded by a Key being of very little value under the fundamental rules, this work does not begin with the first exercises of the *New Higher Arithmetic*.

In most cases, after a short study of an arithmetical problem, the operations themselves, as indicated properly by signs, will suggest the reasoning on which the solution is based. In the solutions here given this has been kept in view, especially where the textbook has presented a *formula*. To the teacher and to the class, alike, it will be advantageous to have the blackboard work written in accordance with this plan; and, on this account, early in the Arithmetic, the subject of Arithmetical Signs has been formally presented.

To read the arithmetical syntax understandingly, and to write it with facility in recording solutions, are acquirements worth far more than they cost. What is here remarked has special reference to the *written work*. Oral explanation, reaching even to particulars, is not to be set aside; on the contrary, the judicious teacher will still require the minute details of an analysis, especially in the examples designated for such exercise in the Arithmetic.

There are a few instances in which the brevity mentioned above has not been observed. The experienced teacher knows, that, in some cases, the operations may be even very few in number and very simple in kind, while the reasons for them are not correspondingly obvious; in such instances, as also where the chief difficulty of the problem is in the complexity of the operation, we have aimed to give an extended solution. Among the articles under which this has been thought advisable, we may mention Compound Subtraction, Proportion, Commission, Stock Investments, Alligation, the applications of Evolution, and the added Miscellany.

Cincinnati, January, 1881

KEY TO

RAY'S NEW HIGHER ARITHMETIC.

MULTIPLICATION.—BILLS AND ACCOUNTS.

Art. 65. (2.)

St. Louis, March 1, 1879.

CHESTER SNYDER,

Bought of Thomas Glenn.

1879.			$			
March	1	4 lb. tea, @ 40 ct. a lb.,	1	60		
"	1	21 " butter, @ 21 ct. a lb.,	4	41		
"	1	58 " bacon, " 13 ct. "	7	54		
"	1	16 " lard, " 9 ct. "	1	44		
"	1	30 " cheese, " 12 ct. "	3	60		
"	1	4 " raisins, " 20 ct. "		80		
"	1	9 doz. eggs, " 15 ct. a doz.,	1	35		
					$20	74

Received payment,

THOMAS GLENN.

Art. 66. (3.)

ALLEGHENY, April 1, 1880.

JAMES WILSON & CO.,

In Acc't with ALLEGHENY COAL CO.

1880.						
		DR.			$	
March	2	To 500 tons coal, @ $2.75 a ton,			1375	00
		CR.	$			
"	3	By 14 bbl. flour, @ $6.55 a bbl.,	91	70		
"	10	" 6123 lb. sugar, @ 8 ct. a lb.,	489	84		
"	15	" cash on acc't,	687	50	1269	04
		Balance due Allegheny Coal Co., - - - - - -			$105	96

CONTRACTIONS IN MULTIPLICATION.

Art. 70. CASE IV.

(1.)

```
  7023
    99
------
702300
  7023
------
695277
```

(2.)

```
   16642
     996
--------
16642000
   66568
--------
16575432
```

(3.)

```
372051000
   744102
---------
372795102
```

Art. 71. CASE V.

(1.)

```
     38057
     48618
----------
  228342
    685026
1826736
----------
1850255226
```

(2.)

```
    267388
     14982
----------
    534776
3743432
  26204024
----------
4006007016
```

(3.)

```
      481063
       63721
-----------
     3367441
     10102323
  30306969
-----------
30653815423
```

$$\begin{array}{r} (4.)\quad 66917 \\ 849612 \\ \hline 803004 \\ 6424032 \\ 5621028 \\ \hline 56853486204 \end{array} \qquad \begin{array}{r} (5.)\quad 102735 \\ 273162 \\ \hline 308205 \\ 2773845 \\ 16643070 \\ \hline 28063298070 \end{array} \qquad \begin{array}{r} (6.)\quad 536712 \\ 729981 \\ \hline 4830408 \\ 43473672 \\ 391263048 \\ \hline 391789562472 \end{array}$$

ARITHMETICAL SIGNS.

Art. 86.

(3.)
$$\left.\begin{array}{rr} 21 \div 3 \times 7 & = +49 \\ -1 \times 1 \div 1 \times 4 \div 2 = & -2 \\ 18 \div 3 \times 6 \div 4 & = +9 \\ 1 \times 4 \times 6 \div 8 & = +3 \end{array}\right\} = 59, \textit{ Ans.}$$

(4.)
$$\left.\begin{array}{r} 16 \times 4 \div 8 = +8 \\ -7 + 48 \div 16 = -4 \\ -3 - 28 \times 0 = -3 \\ 24 \times 6 \div 48 = +3 \\ -4 \times 9 \div 12 = -3 \end{array}\right\} = 1, \textit{ Ans.}$$

(5.)
$$\begin{array}{lr} \left.\begin{array}{lr} 16 \div 16 \times 96 \div 8 & = +12 \\ -7 - 5 + 3 & = -9 \end{array}\right\} & = +3. \\ (27 \div 9) \div 3 - 1 \quad = & 0. \\ 91 \div 13 \times 7 - 45 - 3 = & 1. \end{array}$$

Then, $3 \times 0 + 1 \times 9 = 9$, *Ans.*

CONTRACTIONS IN MULTIPLICATION AND DIVISION.

Art. 88. CASE I.

(1.)
$$\begin{array}{r} 3\underline{)42200} \\ 14066\frac{2}{3}, \textit{ Ans.} \end{array}$$

(2)
$$\begin{array}{r} 656400 \\ 5 \\ \hline 8\underline{)3282000} \\ 410250, \textit{ Ans.} \end{array}$$

(3.)
$$\begin{array}{r} 6\underline{)1072400} \\ 178733\frac{2}{6}, \textit{ Ans.} \end{array}$$

Art. 89. CASE II.

(1.)

4514020000
451402
3)4513568598
1504522866, *Ans.*

(2.)

281257000000
281257
281256718743
5
9)1406283593715
156253732635, *Ans.*

(3.)

6302240000000
630224000
9)6301609776000
700178864000
4
2800715456000, *Ans.*

Art. 90. CASE III.

(1.) 225) 300521761
4 4
9|00) 12020870|44
Quot. 1335652, *Ans.*
$244 \div 4 = 61$, *Rem.*

(2.) 43750) 1510337264
4 4
175000) 6041349056
4 4
7|00000)241653|96224
Quot. = 34521, *Ans.*
$696224 \div 4 \div 4 = 43514$, *Rem.*

(3.) 1406250) 22500712361
8 8
11250000) 180005698888
8 8
9|0000000)144004|5591104
Quot. = 16000, *Ans.*
$45591104 \div 8 \div 8 =$
$712361 =$ *Rem.*

(4.) $20833\frac{1}{3}$) 620712480
3 3
62500)1862137440
4 4
250000)7448549760
4 4
1|000000)29794|199040
Quot. = 29794, *Ans.*
$199040 \div 4 \div 4 \div 3 = 4146\frac{2}{3}$, *Rem.*

(5.) $2916\frac{2}{3})\ 742851692$

$\quad 3 \qquad\qquad 3$

$8750\ )\ 2228555076$

$\quad 8 \qquad\qquad 8$

$7|0000\ )1782844|0608$

Quot. = 254692 *Ans.*

$608 \div 8 \div 3 = 25\frac{1}{3}$, *Rem.*

MISCELLANEOUS EXERCISES.

(1.) $6 × 153 = $918; $918 ÷ 54 = $17, *Ans.*

(2.) 217 × 35 + 25 = 7620, *Ans.*

(3.) 4879 ÷ 41 = 119, *Ans.*

(4.) 103 × 103 = 10609, *Ans.*

(5.) 53815 ÷ 375 = 143 and 190 rem.; 144 times 375 = 54000, this is 185 more than 53815, and since the next lower product is 190 less than 53815, 54000 is the nearest. *Ans.* 54000.

(6.) $2675 ÷ 25 = $107; $107 × 19 = $2033, *Ans.*

(7.) $210 ÷ 15 = $14, gain on each; $75 + $14 = $89, *Ans.*

(8.) 391 mi. — 139 mi. = 252 mi.; 11 hr. — 4 hr. = 7 hr.; 252 mi. ÷ 7 = 36 miles, *Ans.*

(9.) 235 yd. — 12 yd. = 223 yd.; $5 × 235 = $1175; $7 × 223 = $1561; $1561 — $1175 = $386, *Ans.*

(10.) 135 bl. — 83 bl. = 52 bl.; $2 × 52 = $104, *Ans.*

(11.) $75 × 5 = $375; $68 × 12 = $816; $73 × 17 = $1241; $375 + $816 = $1191; $1241 — $1191 = $50 gain, *Ans.*

Also $1191 + $118 = $1309; $1309 ÷ 17 = $77, *Ans.*

(12.) $240 — $24 = $216, whole cost; $216 ÷ 3 = $72, the cost of 1 piece; $72 ÷ $4 = 18. *Ans.* 18 yards.

(14.) 13 men — 8 men = 5 men; 1 man can do it in 13 times 15 days, which is 195 days; and 5 men, in $\frac{1}{5}$ of 195 days, or 39 days, *Ans.*

(15.) 14 men + 7 men = 21 men; 1 man can do it in 14 times 24 days, that is, 336 days; and 21 men, in $\frac{1}{21}$ of 336 days, or 16 days, *Ans.*

(16.) For 1 day, the provisions will support 30 times 45 men, or 1350 men; for 50 days, one fiftieth of 1350 men = 27 men; 45 men — 27 men = 18 men, *Ans.*

(17.) \$18 × 3 = \$54; \$85 + \$54 = \$139, total value; \$139 — \$41 = \$98; 1 sheep cost one fourteenth of \$98, or \$7, *Ans.*

(18.) A sheep and a hog cost \$13; therefore, he will buy as many of each as \$13 is contained times in \$1482; 1482 ÷ 13 = 114, *Ans.*

(19.) 1 horse and 2 oxen cost \$84; therefore, he bought as many horses as \$84 is contained times in \$1260; 1260 ÷ 84 = 15; twice 15 equals 30.

Ans. 15 horses and 30 oxen.

(20.) One seventh of 1050 ct. = 150 ct.; 150 ÷ 25 = 6, of the 25-cent pieces. 1050 ct. — 150 ct. = 900 ct.; 1 of each of the others would make 10 ct. + 5 ct. + 3 ct. = 18 ct.; 900 ÷ 18 = 50, the number of each of the others.

Ans. Of 25-cent pieces, 6; of the others, 50 each.

(21.) \$6300 ÷ 140 = \$45, gain per acre : \$210 — \$45 = \$165, the cost; \$5600 ÷ 140 = \$40, loss per acre; \$165 — \$40 = \$125, sold for per acre.

Ans. \$165, cost; \$125, sold for.

PROPERTIES OF NUMBERS.

LEAST COMMON MULTIPLE.

Art. 104.

(1.)

$$\begin{array}{r|rrr} 2 & 6 & 9 & 20 \\ \hline & \not{3} & 9 & 10 \end{array}$$

2 × 9 × 10 = 180, *Ans.*

(2.)

$$\begin{array}{r|rrr} 2 & \not{15} & 20 & 30 \\ \hline 5 & & 10 & 15 \\ \hline & & 2 & 3 \end{array}$$

2 × 5 × 2 × 3 = 60, *Ans.*

(3.)

7 × 11 × 13 × 5 = 5005, *Ans.*

(4.)

3)~~35~~ 45 63 70
3)15 21 70
~~5~~ ~~7~~ 70

$70 \times 3 \times 3 = 630$, *Ans.*

(5.

2)8 ~~15~~ 20 25 30
2)4 10 25 15
5)2 ~~5~~ 25 15
5 3

$2 \times 2 \times 2 \times 5 \times 5 \times 3 = 600$, *Ans.*

(6.)

3)~~30~~ ~~45~~ 48 80 120 135
5)~~16~~ 80 ~~40~~ 45
16 9

$3 \times 5 \times 16 \times 9 = 2160$, *Ans.*

(7.)

3)174 ~~485~~ 4611 14065 15423
29) 58 1537 14065 5141
97) 2 ~~53~~ 485 5141
See Rem., p. 70, N. H. A. 5 53

$53 \times 5 \times 2 \times 97 \times 29 \times 3 = 4472670$, *Ans.*

(8.)

2)498 85988 235803 490546
83)~~249~~ 42994 235803 245273
7) 518 2841
37) 74 2841 35039
2 2841 ~~947~~

$2841 \times 2 \times 37 \times 7 \times 83 \times 2 = 244291908$, *Ans.*

(9.)

By Art. 100, find the common divisors 37 and 67; then,

37)2183 2479 3953
~~59~~ ~~67~~

$3953 \times 37 = 146261$, *Ans.*

(10.)

Find G. C. D. 31; and again 83. Then,

$$31\overline{)1271 \quad 2573 \quad 3403}$$
$$\cancel{41} \quad \cancel{83}$$

$$3403 \times 31 = 105493,\ \textit{Ans.}$$

Art. 106. CANCELLATION.

(1.) $\dfrac{\overset{10}{\cancel{80}} \times \overset{3}{\cancel{9}}}{\underset{3}{\cancel{24}}} = 30$; hence 30 cows, *Ans.*

(2.) $\dfrac{\overset{4}{\cancel{40}} \times \overset{11}{\cancel{33}} \times \cancel{8}}{\cancel{10} \times \underset{\cancel{3}}{\cancel{24}}} = 4 \times 11 = 44$; hence 44 cents, *Ans.*

(3.) $\dfrac{\overset{\cancel{2}}{\cancel{8}} \times \overset{9}{\cancel{900}} \times \cancel{6}}{\underset{\cancel{4}}{\cancel{400}} \times \underset{\cancel{2}}{\cancel{12}}} = 9$; hence 9 bales, *Ans.*

(4.) $\dfrac{\overset{\cancel{5}}{\cancel{15}} \times \overset{2}{\cancel{24}} \times \overset{2}{\cancel{112}} \times \overset{8}{\cancel{40}} \times \cancel{10}}{\underset{\cancel{5}}{\cancel{25}} \times \underset{\cancel{3}}{\cancel{36}} \times \cancel{56} \times \underset{9}{\cancel{90}}} = \frac{32}{9} = 3\frac{5}{9}$, *Ans.*

COMMON FRACTIONS.—REDUCTION.

Art. 131. CASE I.

(1.)	(2.)	(3.)	(4.)
$15)\frac{30}{45} = \frac{2}{3}$.	$8)\frac{32}{56} = \frac{4}{7}$.	$21)\frac{42}{189} = \frac{2}{9}$.	$15)\frac{105}{195} = \frac{7}{13}$.
(5.)	(6.)	(7.)	(8.)
$14)\frac{154}{210} = \frac{11}{15}$.	$13)\frac{156}{221} = \frac{12}{17}$.	$23)\frac{253}{414} = \frac{11}{18}$.	$29)\frac{667}{783} = \frac{23}{27}$.

(9.)

$93)\frac{1767}{4557} = \frac{19}{49}$, *Ans.*

(10.)

$882)\frac{9702}{18522} = \frac{11}{21}$, *Ans.*

(11.)

$71)\frac{923}{1491} = \frac{13}{21}$, *Ans.*

(12.)

$89)\frac{890}{1691} = \frac{10}{19}$, *Ans.*

(13.)

$133)\frac{2261}{4123} = \frac{17}{31}$, *Ans.*

(14.)

$880)\frac{6160}{40480} = \frac{7}{46}$, *Ans.*

Art. 132. CASE II.

(1.)

$99 \div 11 = 9$;

$\frac{3}{11} \times \frac{9}{9} = \frac{27}{99}$, *Ans.*

$99 \div 33 = 3$;

$\frac{4}{33} \times \frac{3}{3} = \frac{12}{99}$, *Ans.*

(2.)

$63 \div 9 = 7$;

$\frac{4}{9} \times \frac{7}{7} = \frac{28}{63}$, *Ans.*

$63 \div 7 = 9$;

$\frac{3}{7} \times \frac{9}{9} = \frac{27}{63}$, *Ans.*

$63 \div 21 = 3$;

$\frac{3}{21} \times \frac{3}{3} = \frac{9}{63}$, *Ans.*

(3.) $6783 \div 17 = 399$;

$\frac{8}{17} \times \frac{399}{399} = \frac{3192}{6783}$, *Ans.*

$6783 \div 19 = 357$;

$\frac{9}{19} \times \frac{357}{357} = \frac{3213}{6783}$, *Ans.*

$6783 \div 21 = 323$;

$\frac{11}{21} \times \frac{323}{323} = \frac{3553}{6783}$, *Ans.*

Art. 133. CASE III.

(1.) $7 \times 8 + 3 = 59$; *Ans.*, $\frac{59}{8}$.

(2.) $19 \times 4 + 3 = 79$; *Ans.*, $\frac{79}{4}$.

(3.) $13 \times 60 + 37 = 817$; *Ans.*, $\frac{817}{60}$.

(4.) $11 \times 3 + 2 = 35$; *Ans.*, $\frac{35}{3}$.

(5.) $15 \times 11 + 8 = 173$; *Ans.*, $\frac{173}{11}$.

(6.) $127 \times 17 + 11 = 2170$; *Ans.*, $\frac{2170}{17}$.

(7.) $109 \times 19 + 9 = 2080$; *Ans.*, $\frac{2080}{19}$.

(8.) $5 \times 211 + 207 = 1262$; *Ans.*, $\frac{1262}{211}$.

(9.) $13 \times 73 + 51 = 1000$; *Ans.*, $\frac{1000}{73}$.

Art. 134. CASE IV.

(1.) $\$\frac{37}{8} = \$37 \div 8 = \$4\frac{5}{8}$, *Ans.*

(2.) $\frac{137}{4}$ bu. $= 137$ bu. $\div 4 = 34\frac{1}{4}$ bu., *Ans.*

(3.) $\frac{785}{60}$ hr. $= 785$ hr. $\div 60 = 13\frac{1}{12}$ hr., *Ans.*

(5.) $\frac{1295}{37} = 1295 \div 37 = 35$, *Ans.*

(6.) $\frac{800}{9} = 800 \div 9 = 88\frac{8}{9}$, *Ans.*

(7.) $\frac{1162}{11} = 1162 \div 11 = 105\frac{7}{11}$, *Ans.*

(8.) $\frac{4260}{13} = 4260 \div 13 = 327\frac{9}{13}$, *Ans.*

(9.) $\frac{15780}{31} = 15780 \div 31 = 509\frac{1}{31}$, *Ans.*

Art. 135. CASE V.

(1.) $\frac{1 \times \cancel{3} \times \cancel{4}}{\cancel{3} \times \cancel{4} \times 7} = \frac{1}{7}$, *Ans.* (2.) $\frac{\cancel{2} \times \cancel{4} \times \overset{3}{\cancel{21}}}{5 \times \cancel{7} \times \underset{4}{\cancel{8}}} = \frac{3}{5}$, *Ans.*

(3.) $\frac{\cancel{4} \times \overset{\cancel{3}}{\cancel{15}} \times \overset{2}{\cancel{8}}}{\cancel{5} \times \underset{\cancel{4}}{\cancel{16}} \times \cancel{3}} = \frac{2}{1} = 2$, *Ans.*

(4.) $\frac{1 \times \cancel{4} \times \overset{3}{\cancel{15}}}{2 \times \cancel{5} \times \cancel{4}} = \frac{3}{2} = 1\frac{1}{2}$, *Ans.*

$$(5.)\quad \frac{\cancel{3}\times\overset{2}{\cancel{8}}\times\cancel{4}\times\overset{5}{\cancel{35}}}{\cancel{4}\times\underset{3}{\cancel{9}}\times\cancel{7}\times\cancel{4}}=\frac{10}{3}=3\tfrac{1}{3},\ \textit{Ans.}$$

$$(6.)\quad \frac{1\times\cancel{3}\times\overset{\cancel{2}}{\cancel{6}}\times3\times\overset{\cancel{2}}{\cancel{14}}}{\cancel{3}\times5\times\cancel{7}\times\underset{\cancel{2}}{\cancel{4}}\times\cancel{3}}=\tfrac{3}{5},\ \textit{Ans.}$$

$$(7.)\quad \frac{\cancel{8}\times\cancel{3}\times\cancel{4}\times\overset{\cancel{7}}{\cancel{77}}\times\overset{3}{\cancel{57}}}{\cancel{11}\times\cancel{7}\times\cancel{19}\times\underset{\cancel{8}}{\cancel{24}}\times\underset{2}{\cancel{8}}}=\frac{3}{2}=1\tfrac{1}{2},\ \textit{Ans.}$$

$$(8.)\quad \frac{\overset{\cancel{4}}{\cancel{12}}\times\cancel{9}\times\cancel{7}\times\overset{\cancel{5}}{\cancel{10}}\times\overset{3}{\cancel{39}}}{\cancel{13}\times\underset{4}{\cancel{16}}\times\underset{\cancel{2}}{\cancel{18}}\times\underset{\cancel{3}}{\cancel{21}}\times\underset{7}{\cancel{35}}}=\tfrac{3}{28},\ \textit{Ans.}$$

Art. 138. CASE VI.

(1.)

$1 \times 3 \times 5 = 15$
$2 \times 2 \times 5 = 20$
$3 \times 2 \times 3 = 18$
$2 \times 3 \times 5 = 30$ *Den.*

Ans. $\frac{15}{30}$, $\frac{20}{30}$, $\frac{18}{30}$.

(2.)

$1 \times 5 \times 6 = 30$
$1 \times 4 \times 6 = 24$
$1 \times 4 \times 5 = 20$
$4 \times 5 \times 6 = 120$ *Den.*

Ans. $\frac{30}{120}$, $\frac{24}{120}$, $\frac{20}{120}$.

(3.)

$2 \times 7 \times 8 = 112$
$3 \times 3 \times 8 = 72$
$5 \times 3 \times 7 = 105$
$3 \times 7 \times 8 = 168$ *Den.*

Ans. $\frac{112}{168}$, $\frac{72}{168}$, $\frac{105}{168}$.

(4.)

$1 \times 5 \times 6 \times 8 = 240$
$3 \times 2 \times 6 \times 8 = 288$
$5 \times 2 \times 5 \times 8 = 400$
$7 \times 2 \times 5 \times 6 = 420$
$2 \times 5 \times 6 \times 8 = 480$ *Den.*

Ans. $\frac{240}{480}$, $\frac{288}{480}$, $\frac{400}{480}$, $\frac{420}{480}$.

(5.) $\frac{1 \times 7}{2 \times 2} = \frac{7}{4}$;

$\frac{2 \times \not 3}{\not 3 \times 5} = \frac{2}{5}$;

$2 \times 4 \times 5 = 40$
$7 \times 3 \times 5 = 105$
$2 \times 3 \times 4 = 24$
$3 \times 4 \times 5 = 60$ *Den.*

Ans. $\frac{40}{60}$, $\frac{105}{60}$, $\frac{24}{60}$.

(6.) $\frac{2 \times \overset{2}{\not 6}}{\not 3 \times 7} = \frac{4}{7}$; $\frac{\not 3 \times \overset{2}{\not 8}}{\not 4 \times \underset{3}{\not 9}} = \frac{2}{3}$;

$\frac{1 \times \not 4 \times 3 \times \overset{3}{\not{21}}}{2 \times 5 \times \not 7 \times \underset{2}{\not 8}} = \frac{9}{20}$.

$4 \times 3 \times 20 = 240$
$2 \times 7 \times 20 = 280$
$9 \times 3 \times 7 = 189$
$7 \times 3 \times 20 = 420$ *Den.*

Ans. $\frac{240}{420}$, $\frac{280}{420}$, $\frac{189}{420}$.

(1.) $\frac{1 \times 4}{2 \times 4} = \frac{4}{8}$

$\frac{3 \times 2}{4 \times 2} = \frac{6}{8}$.

Ans. $\frac{4}{8}$, $\frac{6}{8}$, $\frac{5}{8}$.

(2.) $\frac{2 \times 4}{3 \times 4} = \frac{8}{12}$

$\frac{5 \times 2}{6 \times 2} = \frac{10}{12}$.

Ans. $\frac{8}{12}$, $\frac{10}{12}$, $\frac{7}{12}$.

(3.) $\frac{3 \times 5}{4 \times 5} = \frac{15}{20}$

$\frac{4 \times 4}{5 \times 4} = \frac{16}{20}$

$\frac{9 \times 2}{10 \times 2} = \frac{18}{20}$.

Ans. $\frac{15}{20}$, $\frac{16}{20}$, $\frac{18}{20}$, $\frac{11}{20}$.

Art. 139. CASE VII.

(1.) The L. C. M. of 3, 4, and $6 = 12$:

$\frac{1 \times 4}{3 \times 4} = \frac{4}{12}$; $\frac{3 \times 3}{4 \times 3} = \frac{9}{12}$; $\frac{5 \times 2}{6 \times 2} = \frac{10}{12}$. *Ans.* $\frac{4}{12}$, $\frac{9}{12}$, $\frac{10}{12}$.

(2.) The L. C. M. of 2, 5, 10, and 4 is 20:

$\frac{1 \times 10}{2 \times 10} = \frac{10}{20}$; $\frac{3 \times 4}{5 \times 4} = \frac{12}{20}$; $\frac{9 \times 2}{10 \times 2} = \frac{18}{20}$; $\frac{3 \times 5}{4 \times 5} = \frac{15}{20}$.

Ans. $\frac{10}{20}$, $\frac{12}{20}$, $\frac{18}{20}$, $\frac{15}{20}$.

(3.) The L. C. M. of 7, 8, and 14 is 56:

$\frac{3 \times 8}{7 \times 8} = \frac{24}{56}$; $\frac{5 \times 7}{8 \times 7} = \frac{35}{56}$; $\frac{11 \times 4}{14 \times 4} = \frac{44}{56}$. *Ans.* $\frac{24}{56}, \frac{35}{56}, \frac{44}{56}$.

(4.) $2)\frac{6}{8} = \frac{3}{4}$; $3)\frac{9}{12} = \frac{3}{4}$; $5)\frac{15}{20} = \frac{3}{4}$. *Ans.* $\frac{3}{4}, \frac{3}{4}, \frac{3}{4}, \frac{3}{4}$.

(5.) $3)\frac{6}{9} = \frac{2}{3}$; $3)\frac{9}{12} = \frac{3}{4}$; $4)\frac{12}{20} = \frac{3}{5}$;

L. C. M. of 3, 4, 5, and 10, is 60: $\frac{2 \times 20}{3 \times 20} = \frac{40}{60}$;

$\frac{3 \times 15}{4 \times 15} = \frac{45}{60}$; $\frac{3 \times 12}{5 \times 12} = \frac{36}{60}$; $\frac{7 \times 6}{10 \times 6} = \frac{42}{60}$.

Ans. $\frac{40}{60}, \frac{45}{60}, \frac{36}{60}, \frac{42}{60}$.

(6.) $1\frac{3}{4} = \frac{7}{4}$; $3\frac{2}{3} = \frac{11}{3}$; $\frac{3 \times \overset{5}{\cancel{25}}}{\underset{2}{\cancel{10}} \times 7} = \frac{15}{14}$;

L. C. M. of 4, 3, and 14, is 84; $\frac{7 \times 21}{4 \times 21} = \frac{147}{84}$;

$\frac{11 \times 28}{3 \times 28} = \frac{308}{84}$; $\frac{15 \times 6}{14 \times 6} = \frac{90}{84}$. *Ans.* $\frac{147}{84}, \frac{308}{84}, \frac{90}{84}$.

ADDITION OF FRACTIONS.

Art. 140.

(2.). L. C. M. of 3, 5, 9, and 15, is 45:

	45
3	15 × 2 = 30
5	9 × 3 = 27
9	5 × 7 = 35
15	3 × 4 = 12
	104; $\frac{104}{45} = 2\frac{14}{45}$, *Ans.*

(3.) $1\frac{2}{3} = \frac{5}{3}$; $2\frac{3}{5} = \frac{13}{5}$:

	15
3	5 × 5 = 25
5	3 × 13 = 39
	64

$\frac{64}{15} = 4\frac{4}{15}$, *Ans.*

(4.) $2\frac{1}{4}=\frac{9}{4}$; $3\frac{2}{7}=\frac{23}{7}$; $4\frac{5}{6}=\frac{29}{6}$; L. C. M. of 4, 7, and 6, is 84:

	84
4	$21 \times 9 = 189$
7	$12 \times 23 = 276$
6	$14 \times 29 = 406$
	871

$\frac{871}{84}=10\frac{31}{84}$, *Ans.*

(5.) $\frac{6}{10}=\frac{3}{5}$; $\frac{4}{14}=\frac{2}{7}$; $\frac{8}{12}=\frac{2}{3}$; $2\frac{1}{3}=\frac{7}{3}$; L. C. M. of 5, 7, 3, and 3, is 105:

	105
5	$21 \times 3 = 63$
7	$15 \times 2 = 30$
3	$35 \times 2 = 70$
3	$35 \times 7 = 245$
	408

$\frac{408}{105}=3\frac{93}{105}=3\frac{31}{35}$, *Ans.*

(6.) $1+2+3+4=10$; L. C. M. of 2, 3, 4, and 5, is 60:

	60
2	30
3	20
4	15
5	12
	77

$\frac{77}{60}=1\frac{17}{60}$;

$10+1\frac{17}{60}=11\frac{17}{60}$, *Ans.*

(7.) $\frac{2}{3}$ of $\frac{4}{5}=\frac{8}{15}$;

$\frac{\not{3}}{\not{7}}$ of $\frac{5}{8}$ of $\frac{\not{7}}{\not{3}}=\frac{5}{8}$;

L. C. M. of 8 and 15, is 120:

	120
15	$8 \times 8 = 64$
8	$15 \times 5 = 75$
	139

$\frac{139}{120}=1\frac{19}{120}$, *Ans.*

(8.) L. C. M. of 8, 22, 24, and 88, is 264:

	264
8	$33 \times 3 = 99$
22	$12 \times 1 = 12$
24	$11 \times 7 = 77$
88	$3 \times 29 = 87$
	275

$\frac{275}{264}=1\frac{1}{24}$, *Ans.*

(9.) L. C. M. of 8, 12, 18, 24, and 27, is 216:

	216
8	$27 \times 7 = 189$
12	$18 \times 11 = 198$
18	$12 \times 17 = 204$
24	$9 \times 23 = 207$
27	$8 \times 20 = 160$
	958

$\frac{958}{216}=4\frac{94}{216}=4\frac{47}{108}$, *Ans.*

(10.) $96\frac{1}{4} = \frac{385}{4}$:

$$\frac{\cancel{4}}{\cancel{7}} \times \frac{\overset{55}{\cancel{385}}}{\cancel{4}} = 55;$$

$$\frac{\overset{2}{\cancel{8}}}{9} \times \frac{11}{\underset{3}{\cancel{12}}} \times \frac{31}{\underset{3}{\cancel{6}}} = \frac{341}{81} = 4\frac{17}{81};$$

$55 + 4\frac{17}{81} = 59\frac{17}{81}$, *Ans.*

(11.) L. C. M. of denominators is 729:

	729	
3	243 ×	2 = 486
9	81 ×	8 = 648
27	27 ×	26 = 702
81	9 ×	80 = 720
243	3 ×	242 = 726
		728
		4010

$\frac{4010}{729} = 5\frac{365}{729}$, *Ans.*

SUBTRACTION OF FRACTIONS.

Art. 141.

(1.) $\frac{4}{5} - \frac{5}{12} = \frac{48}{60} - \frac{25}{60} = \frac{23}{60}$, *Ans.*

(2.) $\frac{3}{17}$ of $\frac{1}{2} = \frac{3}{34}$; $\frac{8}{11} - \frac{3}{34} = \frac{272}{374} - \frac{33}{374} = \frac{239}{374}$, *Ans.*

(3.) $\frac{3}{28}$ of $\frac{5}{6} = \frac{5}{56}$; $\frac{11}{54} - \frac{5}{56} = \frac{308}{1512} - \frac{135}{1512} = \frac{173}{1512}$, *Ans.*

(4.) $\frac{1}{13}$ of $\frac{4}{1} = \frac{4}{13}$; $\frac{5}{11} - \frac{4}{13} = \frac{65}{143} - \frac{44}{143} = \frac{21}{143}$, *Ans.*

(5.) $\frac{11}{14} - \frac{4}{63} = \frac{99}{126} - \frac{8}{126} = \frac{91}{126} = \frac{13}{18}$, *Ans.*

(6.) $\frac{9}{55} - \frac{1}{15} = \frac{27}{165} - \frac{11}{165} = \frac{16}{165}$, *Ans.*

(7.) $\frac{7}{45} - \frac{11}{75} = \frac{35}{225} - \frac{33}{225} = \frac{2}{225}$, *Ans.*

(8.) $\frac{10}{39} - \frac{8}{65} = \frac{50}{195} - \frac{24}{195} = \frac{26}{195} = \frac{2}{15}$, *Ans.*

(9.) $12\frac{3}{4} - 10\frac{13}{16} = \frac{51}{4} - \frac{173}{16} = \frac{204}{16} - \frac{173}{16} = \frac{31}{16} = 1\frac{15}{16}$, *Ans.*

(10.) $12 - 9 = 3$; $\frac{23}{28} - \frac{27}{35} = \frac{115}{140} - \frac{108}{140} = \frac{7}{140} = \frac{1}{20}$; $3\frac{1}{20}$, *Ans.*

(11.) $5 - 2 = 3$; $\frac{23}{32} - \frac{2}{7} = \frac{161}{224} - \frac{64}{224} = \frac{97}{224}$; $3\frac{97}{224}$, *Ans.*

(12.) $7\frac{5}{12} - 3\frac{1}{2} = \frac{89}{12} - \frac{42}{12} = \frac{47}{12} = 3\frac{11}{12}$, *Ans.*

(13.) $1 = \frac{7}{7}$; $\frac{7}{7} - \frac{3}{7} = \frac{4}{7}$; $15 - \frac{3}{7} = 14\frac{4}{7}$, *Ans.*

(14.) $1 - \frac{3}{8} = \frac{5}{8}$; $18 - 5\frac{3}{8} = 12\frac{5}{8}$, *Ans.*

(15.) $2\frac{7}{9} = \frac{25}{9}$; $\frac{5}{3}$ of $\frac{25}{9} = \frac{125}{27}$; $\frac{125}{27} - \frac{71}{18} = \frac{250}{54} - \frac{213}{54} = \frac{37}{54}$, *Ans.*

(16.) $\frac{4}{5}$ of $1\frac{7}{8} = \frac{3}{2}$; $3\frac{1}{3} - \frac{3}{2} = \frac{20}{6} - \frac{9}{6} = \frac{11}{6} = 1\frac{5}{6}$, *Ans.*

(17.) $\frac{16}{3}$ of $4\frac{1}{2} = 24$; $\frac{13}{4}$ of $\frac{16}{5} = \frac{52}{5}$; $\frac{24}{1} - \frac{52}{5} = \frac{120}{5} - \frac{52}{5} = \frac{68}{5} = 13\frac{3}{5}$, *Ans.*

(18.) $11 + 8 - 9 = 10$; $\frac{2}{3} + \frac{7}{9} - \frac{19}{22} = \frac{132}{198} + \frac{154}{198} - \frac{171}{198} = \frac{115}{198}$. *Ans.* $10\frac{115}{198}$.

(19.) $\frac{3}{5}$ of $\frac{25}{38} = \frac{15}{38}$; $\frac{25}{38} - \frac{15}{38} = \frac{10}{38} = \frac{5}{19}$, *Ans.*

(20.) $\frac{4}{7}$ of $\frac{5}{8} = \frac{5}{14}$; $\frac{1}{5}$ of $\frac{3}{7} = \frac{3}{35}$; $\frac{5}{14} + \frac{3}{35} = \frac{25}{70} + \frac{6}{70} = \frac{31}{70}$; $\frac{70}{70} - \frac{31}{70} = \frac{39}{70}$, *Ans.*

(21.) $\frac{13}{4} + \frac{22}{5} - \frac{11}{2} + \frac{133}{8} - \frac{179}{24} + \frac{10}{1} - \frac{89}{6} = \frac{390}{120} + \frac{528}{120} - \frac{660}{120} + \frac{1995}{120} - \frac{895}{120} + \frac{1200}{120} - \frac{1780}{120} = \frac{778}{120} = 6\frac{58}{120} = 6\frac{29}{60}$, *Ans.*

(22.) $\frac{26}{5} - \frac{17}{6} + \frac{13}{2} - \frac{33}{10} + \frac{37}{12} + \frac{73}{9} - \frac{65}{4} = \frac{936}{180} - \frac{510}{180} + \frac{1170}{180} - \frac{594}{180} + \frac{555}{180} + \frac{1460}{180} - \frac{2925}{180} = \frac{92}{180} = \frac{23}{45}$, *Ans.*

(23.) $\frac{3}{8}$ of $\frac{5}{6} = \frac{5}{16}$; $\frac{2}{3}$ of $\frac{3}{7} = \frac{2}{7}$; $1 - \frac{5}{16} - \frac{2}{7} = \frac{112}{112} - \frac{35}{112} - \frac{32}{112} = \frac{45}{112}$, *Ans.*

MULTIPLICATION OF FRACTIONS.

Art. 142.

(1.) $\frac{10}{13} \times 12 = \frac{120}{13} = 9\frac{3}{13}$, *Ans.*

(2.) $\dfrac{11}{\underset{4}{\cancel{24}}} \times \dfrac{\overset{3}{\cancel{18}}}{1} = \dfrac{33}{4} = 8\frac{1}{4}$, *Ans.*

(3.) $\frac{29}{\underset{2}{\cancel{48}}} \times \frac{\cancel{24}}{1} = \frac{29}{2} = 14\frac{1}{2}$, *Ans.*

(4.) $\frac{9}{\underset{4}{\cancel{16}}} \times \frac{\overset{7}{\cancel{28}}}{1} = \frac{63}{4} = 15\frac{3}{4}$, *Ans.*

(5.) $\frac{13}{\cancel{15}} \times \frac{\overset{2}{\cancel{30}}}{1} = 26$, *Ans.*

(7.) $\frac{\overset{5}{\cancel{45}}}{1} \times \frac{7}{\cancel{9}} = 35$, *Ans.*

(8.) $\frac{\overset{25}{\cancel{50}}}{1} \times \frac{11}{\underset{7}{\cancel{14}}} = \frac{275}{7} = 39\frac{2}{7}$, *Ans.*

(9.) $\frac{25}{1} \times \frac{3}{4} = \frac{75}{4} = 18\frac{3}{4}$, *Ans.*

(10.) $\frac{\overset{4}{\cancel{32}}}{1} \times \frac{19}{\cancel{8}} = \frac{76}{1} = 76$, *Ans.*

(11.) $\frac{28}{1} \times \frac{11}{3} = \frac{308}{3} = 102\frac{2}{3}$, *Ans.*

(12.) $\frac{\overset{3}{\cancel{12}}}{\underset{5}{\cancel{35}}} \times \frac{\cancel{7}}{\underset{4}{\cancel{16}}} = \frac{3}{20}$, *Ans.*

(13.) $\frac{\overset{3}{\cancel{15}}}{\underset{2}{\cancel{16}}} \times \frac{\overset{3}{\cancel{24}}}{\underset{5}{\cancel{25}}} = \frac{9}{10}$, *Ans.*

(14.) $\frac{\overset{6}{\cancel{42}}}{\underset{5}{\cancel{55}}} \times \frac{\overset{2}{\cancel{22}}}{\underset{5}{\cancel{35}}} = \frac{12}{25}$, *Ans.*

(16.) $6\frac{2}{3} = \frac{20}{3}$; $4\frac{1}{2} = \frac{9}{2}$; $\frac{\overset{10}{\cancel{20}}}{\cancel{3}} \times \frac{\overset{3}{\cancel{9}}}{\cancel{2}} = 30$, *Ans.*

(17.) $4\frac{4}{5} = \frac{24}{5}$; $2\frac{2}{3} = \frac{8}{3}$; $\frac{\overset{8}{\cancel{24}}}{5} \times \frac{8}{\cancel{3}} = \frac{64}{5} = 12\frac{4}{5}$, *Ans.*

(18.) $\dfrac{\overset{9}{\cancel{99}}}{\underset{2}{\cancel{8}}} \times \dfrac{\overset{9}{\cancel{36}}}{\cancel{11}} = \dfrac{81}{2} = 40\tfrac{1}{2}$, *Ans.*

(19.) $\dfrac{\overset{5}{\cancel{95}}}{\underset{3}{\cancel{12}}} \times \dfrac{\overset{16}{\cancel{64}}}{\cancel{19}} = \dfrac{80}{3} = 26\tfrac{2}{3}$, *Ans.*

(20.) $\frac{1 \times 8}{5 \times 1} \times \frac{1 \times 10}{4 \times 1} = 2 \times 2 = 4$, *Ans.*

(21.) $\frac{2 \times 27}{3 \times 5} \times \frac{3 \times 10}{5 \times 3} = \frac{2 \times 9 \times 2}{5} = \frac{36}{5} = 7\frac{1}{5}$, *Ans.*

(22.) $\frac{3 \times 2 \times 28}{4 \times 3 \times 5} \times \frac{3 \times 27}{7 \times 8} = \frac{3 \times 27}{5 \times 4} = \frac{81}{20} = 4\frac{1}{20}$, *Ans.*

(23.) $\frac{5}{1} \times \frac{17}{4} \times \frac{7}{3} \times \frac{3 \times 40}{7 \times 9} = \frac{5 \times 17 \times 10}{9} = \frac{850}{9} = 94\frac{4}{9}$, *Ans.*

(24.) $\frac{4}{5} \times \frac{3}{7} \times \frac{5}{11} \times \frac{1 \times 5}{3 \times 2} \times \frac{4 \times 22}{7 \times 7} = \frac{4 \times 4 \times 5}{7 \times 7 \times 7} = \frac{80}{343}$, *Ans.*

(25.) $\frac{3}{5} \times \frac{7}{9} \times \frac{9}{11} \times \frac{10}{3} \times \frac{22}{7} = \frac{2 \times 2}{1} = 4$, *Ans.*

(26.) $\frac{7}{2} \times \frac{14}{3} \times \frac{28}{5} \times \frac{2 \times 5}{9 \times 14} \times \frac{27}{4} = \frac{7 \times 7}{1} = 49$, *Ans.*

(27.) $\frac{7}{8} \times \frac{25}{1} = \frac{175}{8} = 21\frac{7}{8}$; hence \$$21\frac{7}{8}$, *Ans.*

(28.) $\frac{51}{4} \times \frac{25}{1} = \frac{1275}{4} = 318\frac{3}{4}$; hence $318\frac{3}{4}$ days, *Ans.*

(29.) $\frac{7}{2} \times \frac{11}{4} = \frac{77}{8} = 9\frac{5}{8}$; hence $9\frac{5}{8}$ cents, *Ans.*

(30.) $\frac{3 \times 2}{5 \times 3} \times \frac{75}{4} = \frac{15}{2} = 7\frac{1}{2}$; hence \$$7\frac{1}{2}$, *Ans.*

(31.) $\frac{4}{4} - \frac{3}{4} = \frac{1}{4}$; $\frac{5}{8} \times \frac{1}{4} = \frac{5}{32}$, *Ans.*

REMARK.—The examples under this article, as also those of the following, are comparatively easy, and hence the solutions are briefly indicated here. The pupils may work with like brevity, when the multiplications or divisions *are mere parts* of a solution requiring *various* processes. But where, as in these articles, the multiplication or division *itself is* the solution *to be exhibited*, the pupil should have frequent exercise in writing, with particularity and accuracy, the very statements and the steps which follow. Thus, in the next article, take the 6th and the 15th, for illustrations.

DIVISION OF FRACTIONS.

Art. 143.

(6.) $\frac{3}{4} \div \frac{1}{2} = \frac{3}{\not{4}_2} \times \frac{\not{2}}{1} = \frac{3}{2} = 1\frac{1}{2}$, *Ans.*

(15.) $1\frac{1}{2} = \frac{3}{2}$;
$\frac{1}{2}$ of $\frac{3}{5}$ of $7\frac{1}{2} = \frac{3}{10}$ of $\frac{15}{2} = \frac{9}{4}$;
$\frac{3}{2} \div \frac{9}{4} = \frac{3}{2} \times \frac{4}{9} = \frac{2}{3}$, *Ans.*

Much may be gained by practice in the proper use of signs and punctuation marks. Accuracy in the use of arithmetical syntax, is worth all the time it may cost the class to make the acquirement.

(1.)
$\frac{9}{16} \times \frac{1}{3} = \frac{3}{16}$, *Ans.*

(2.)
$\frac{14}{23} \times \frac{1}{7} = \frac{2}{23}$, *Ans.*

(3.)
$\frac{3}{5} \times \frac{1}{8} = \frac{3}{40}$, *Ans.*

(4.)
$\frac{6}{1} \times \frac{3}{2} = 9$, *Ans.*

(5.)
$\frac{21}{1} \times \frac{10}{9} = 23\frac{1}{3}$, *Ans.*

(7.)
$\frac{2}{3} \times \frac{40}{1} = 26\frac{2}{3}$, *Ans.*

(8.)
$\frac{21}{25} \times \frac{15}{14} = \frac{3 \times 3}{5 \times 2} = \frac{9}{10}$, *Ans.*

(9.)
$\frac{12}{35} \times \frac{77}{30} = \frac{2 \times 11}{5 \times 5} = \frac{22}{25}$, *Ans.*

(10.)
$\frac{7}{4} \times \frac{1}{5} = \frac{7}{20}$, *Ans.*

(11.)
$\frac{49}{6} \times \frac{3}{2} = 12\frac{1}{4}$, *Ans.*

(12.)
$\frac{39}{2} \times \frac{8}{15} = \frac{52}{5} = 10\frac{2}{5}$, *Ans.*

(13.)
$\frac{147}{2} \times \frac{5}{49} = \frac{3 \times 5}{2} = 7\frac{1}{2}$, *Ans.*

(14.) $54\frac{43}{48} = \frac{2635}{48}$; $25\frac{5}{6} = \frac{155}{6}$; $\frac{2635}{48} \times \frac{6}{155} = \frac{17}{8} = 2\frac{1}{8}$, *Ans.*

Remark.—We have before mentioned that usage is not well settled in regard to the effect of the signs ÷ and ×, occurring in succession. In fractions, we seem to have a facility for avoiding ambiguity, which usage has not afforded in whole numbers. The word "of" is the convenient connective by which the divisor is shown to be a *result*, and not a *mere factor* of a product. Thus, in $\frac{3}{21} \div \frac{4}{5}$ of $\frac{3}{8}$, the operator, recognizing the compound fraction, is not likely to use $\frac{4}{5}$ as the divisor; it seems most natural to use $\frac{3}{10}$.

Where the divisor is a compound fraction, the operator may save time and space by making *all* the inversions *at once*, as we illustrate in examples 24th and 25th following.

(15.) See solution on page 23.

(16.) $\frac{3}{10}$ of $\frac{3}{16}$ of $\frac{5}{12} = \frac{3}{128}$; $\frac{7}{24}$ of $\frac{12}{35} = \frac{1}{10}$; $\frac{3}{128} \times 10 = \frac{15}{64}$, *Ans.*

(17.) $\frac{3}{4}$ of $\frac{1}{2} = \frac{3}{8}$; $\frac{1}{4}$ of $\frac{2}{3} = \frac{1}{6}$; $\frac{3}{8} \div \frac{1}{6} = 2\frac{1}{4}$, *Ans.*

(18.) $\frac{7}{9}$ of $\frac{18}{5} = \frac{14}{5}$; $\frac{13}{14}$ of $7 = \frac{13}{2}$; $\frac{14}{5} \times \frac{2}{13} = \frac{28}{65}$, *Ans.*

(19.) $\frac{1}{3}$ of $\frac{4}{5} \times \frac{39}{320} = \frac{13}{400}$; $\frac{2}{75}$ of $\frac{13}{4} = \frac{13}{150}$; $\frac{13}{400} \times \frac{150}{13} = \frac{3}{8}$, *Ans.*

(20.) $\frac{2}{7}$ of $\frac{11}{2} = \frac{11}{7}$; $\frac{2}{5}$ of $\frac{9}{14}$ of $\frac{10}{3} = \frac{6}{7}$; $\frac{11}{7} \times \frac{7}{6} = 1\frac{5}{6}$, *Ans.*

(21.) $\frac{1}{3}$ of $\frac{2}{7}$ of $\frac{4}{11} = \frac{8}{231}$; $\frac{2}{5}$ of $\frac{1}{3}$ of $\frac{4}{7} = \frac{8}{105}$; $\frac{8}{231} \times \frac{105}{8} = \frac{5}{11}$, *Ans.*

(22.) $\frac{15}{8} \times \frac{14}{3} \times \frac{11}{18} \times \frac{9}{35} = 1\frac{3}{8}$, *Ans.* See Remark.

(23.) $\frac{22}{7} \times \frac{9}{4} \times \frac{4}{33} \times \frac{11}{5} \times \frac{10}{33} = \frac{4}{7}$, *Ans.*

(24.) $\frac{9}{11} \times \frac{2}{3} \times \frac{55}{2} \times \frac{9}{4} \times \frac{17}{3} \times \frac{2}{11} = \frac{765}{22} = 34\frac{17}{22}$, *Ans.*

(25.) $\frac{21}{8} \times \frac{4}{5} \times \frac{58}{3} \times \frac{6}{29} \times \frac{10}{3} \times \frac{1}{8} = \frac{7}{2} = 3\frac{1}{2}$, *Ans.*

COMPLEX FRACTIONS.

Art. 144.

(1.) $\frac{3}{8} \div \frac{11}{4} = \frac{3}{8} \times \frac{4}{11} = \frac{3}{22}$, *Ans.*

(2.) $\frac{7}{4} \div \frac{14}{5} = \frac{7}{4} \times \frac{5}{14} = \frac{5}{8}$, *Ans.*

(3.) $\frac{14}{3} \div \frac{35}{12} = \frac{14}{3} \times \frac{12}{35} = \frac{8}{5} = 1\frac{3}{5}$, *Ans.*

(4.) $\frac{33}{14} \div \frac{44}{21} = \frac{33}{14} \times \frac{21}{44} = \frac{9}{8} = 1\frac{1}{8}$, *Ans.*

(5.) $\frac{99}{8} \div 18 = \frac{99}{8 \times 18} = \frac{11}{16}$, *Ans.*

(6.) $62 \div \frac{186}{11} = \frac{62 \times 11}{186} = \frac{11}{3} = 3\frac{2}{3}$, *Ans.*

(7.) $4\frac{1}{4} \div 8 = \frac{17}{4 \times 8} = \frac{17}{32}$; $\frac{7}{10} \times \frac{17}{32} = \frac{119}{320}$, *Ans.*

(8.) $\frac{17}{2} \div \frac{51}{5} = \frac{17}{2} \times \frac{5}{51} = \frac{5}{6}$; $\frac{31}{25} \times \frac{5}{6} = \frac{31}{30} = 1\frac{1}{30}$, *Ans.*

(9.) $\frac{16}{3} \div \frac{46}{5} = \frac{16}{3} \times \frac{5}{46} = \frac{40}{69}$; $\frac{143}{12} \div \frac{121}{5} = \frac{143}{12} \times \frac{5}{121} = \frac{65}{132}$; $\frac{40}{69} \times \frac{65}{132} = \frac{650}{2277}$, *Ans.*

(10.) $\frac{25}{4} \times \frac{5}{12} \times \frac{25}{2} \times \frac{1}{7} = \frac{3125}{672} = 4\frac{437}{672}$, *Ans.*

(11.) $\frac{52}{7} \times \frac{9}{365} \times \frac{73}{1} \times \frac{3}{52} = \frac{9 \times 3}{7 \times 5} = \frac{27}{35}$, *Ans.*

(12.) $\frac{26}{11} \times \frac{5}{13} \times \frac{87}{10} \times \frac{11}{29} = 3$, *Ans.*

GREATEST COMMON DIVISOR OF FRACTIONS.

Art. 145.

(1.) $83\frac{1}{3} = \frac{250}{3}$; $268\frac{3}{4} = \frac{1075}{4}$; G. C. D. of 250 and 1075, is 25; L. C. M. of 3 and 4, is 12. *Ans.* $\frac{25}{12} = 2\frac{1}{12}$.

(2.) $14\frac{7}{12} = \frac{175}{12}$; $95\frac{3}{8} = \frac{763}{8}$; G. C. D. of numerators $= 7$; L. C. M. of denominators $= 24$. *Ans.* $\frac{7}{24}$.

(3.) $59\frac{1}{9} = \frac{532}{9}$; $735\frac{14}{15} = \frac{11039}{15}$; G. C. D. of numerators $= 133$; L. C. M. of denominators $= 45$.

Ans. $\frac{133}{45} = 2\frac{43}{45}$.

(4.) $23\frac{7}{16}=\frac{375}{16}$; $213\frac{13}{24}=\frac{5125}{24}$; G. C. D. of numerators $=125$; L. C. M. of denominators $=48$.

Ans. $\frac{125}{48}=2\frac{29}{48}$.

(5.) $418\frac{3}{5}=\frac{2093}{5}$; $1772\frac{1}{3}=\frac{5317}{3}$; G. C. D. of numerators $=13$; L. C. M. of denominators $=15$. *Ans.* $\frac{13}{15}$.

(6.) $261\frac{13}{14}=\frac{3667}{14}$; $652\frac{11}{21}=\frac{13703}{21}$; G. C. D. of numerators $=193$; L. C. M. of denominators $=42$.

Ans. $\frac{193}{42}=4\frac{25}{42}$.

(7.) $44\frac{4}{9}=\frac{400}{9}$; $546\frac{2}{3}=\frac{1640}{3}$; $3160=\frac{3160}{1}$; G. C. D. of numerators $=40$; L. C. M. of denominators $=9$.

Ans. $\frac{40}{9}=4\frac{4}{9}$.

(8.) The quantities are $\frac{275}{2}$ bu., $\frac{3825}{8}$ bu., $\frac{8375}{4}$ bu. The greatest common measure of these is the capacity of sack required. G. C. D. of numerators $=25$; L. C. M. of denominators $=8$; hence $\frac{25}{8}$, or $3\frac{1}{8}$, is the number of bushels in each sack, *Ans.*

Dividing each quantity by $3\frac{1}{8}$ we have, also, 44, 153, 670, *Ans.*

(9.) The sides are $\frac{539}{4}$ ft., $\frac{385}{3}$ ft., $\frac{231}{2}$ ft. The greatest number of feet which will exactly measure these sides, is the G. C. D. of these numbers. G. C. D. of numerators $=77$, and L. C. M. of denominators $=12$; hence $\frac{77}{12}$ ft. is the *necessary* length of the equal rails. The number in one round is $(\frac{539}{4}+\frac{385}{3}+\frac{231}{2})\div\frac{77}{12}=59$; hence, in 6 rounds, 354 rails, *Ans.*

REMARK.—$\frac{77}{12}$ being the necessary reach, $\frac{77}{12}$ ft. $+\frac{6}{12}$ ft., or, 6 ft. 11 in. is the *allowed* length of rails.

LEAST COMMON MULTIPLE OF FRACTIONS.

Art. 146.

(1.) L. C. M. of numerators $=60$; G. C. D. of denominators $=1$. *Ans.* $\frac{60}{1}=60$.

(2.) The numbers are $\frac{9}{2}$, $\frac{27}{4}$, $\frac{35}{6}$, and $\frac{21}{2}$; L. C. M. of numerators $= 945$; G. C. D. of denominators $= 2$.

Ans. $\frac{945}{2} = 472\frac{1}{2}$.

(3.) Numbers $= \frac{10}{3}, \frac{35}{8}, \frac{5}{12}, \frac{50}{9}, \frac{25}{2}$; L. C. M. of numerators $= 350$; G. C. D. of denominators $= 1$.

Ans. $\frac{350}{1} = 350$.

(4.) Their times are $\frac{100}{7}$, $\frac{100}{11}$, $\frac{50}{3}$, $\frac{25}{1}$, in hours, respectively. Any common multiple of these will express the time to elapse till they are all together, and the *least* common multiple will express the time to elapse till they *first* come together. By the Rule this is found, G. C. D. of numerators $\div$ L. C. M. of denominators $= 100 \div 1$; hence 100 hr., *Ans.*

PROMISCUOUS EXERCISES.

(1.) $\frac{7}{2} + \frac{13}{3} + \frac{21}{4} + \frac{21}{32} + \frac{5}{48} = \frac{336}{96} + \frac{416}{96} + \frac{504}{96} + \frac{63}{96} + \frac{10}{96} = \frac{1329}{96} = 13\frac{27}{32}$, *Ans.*

(2.) $1\frac{1}{26} = \frac{27}{26}$; $1 \div 1\frac{4}{9} = \frac{18}{26}$; sum $= \frac{45}{26}$, difference $= \frac{9}{26}$; and 45 contains 9, *five* times, *Ans.*

(3.) $\frac{11}{4} + \frac{5}{2}$ of $(7 \div \frac{19}{5}) - (\frac{5}{3} \div \frac{5}{2}) = \frac{11}{4} + \frac{5}{2}$ of $\frac{35}{19} - \frac{2}{3} = \frac{11}{4} + \frac{175}{38} - \frac{2}{3} = \frac{627+1050-152}{228} = \frac{1525}{228}$; $\frac{1525}{228} \div \frac{305}{228} = 5$, *Ans.*

(4.) $4\frac{4}{15} \times 2\frac{5}{8} = \frac{64}{15} \times \frac{21}{8} = \frac{56}{5}$; $5\frac{1}{5} - 4\frac{1}{2} = \frac{26}{5} - \frac{9}{2} = \frac{7}{10}$; $\frac{56}{5} \div \frac{7}{10} = 16$, *Ans.*

Also, $100 - \frac{200}{3} + (\frac{22}{3} \div \frac{9}{4}) = \frac{100}{3} + \frac{88}{27} = \frac{988}{27}$; $\frac{988}{27} \times \frac{5}{7} = \frac{4940}{189} = 26\frac{26}{189}$, *Ans.*

(5.) $\frac{1}{4}$ of $\frac{21}{4} - \frac{1}{3}$ of $\frac{31}{8} = \frac{21}{16} - \frac{31}{24} = \frac{63-62}{48} = \frac{1}{48}$, *Ans.*

(6.) $\frac{32}{51} \times \frac{85}{112} \times \frac{189}{207} \times \frac{23}{36} = \frac{32}{112} \times \frac{85}{51} \times \frac{23}{207} \times \frac{189}{36} = \frac{2}{7} \times \frac{5}{3} \times \frac{1}{9} \times \frac{21}{4} = \frac{5}{18}$, *Ans.*

(7.) $\frac{1}{2}$ of $(\frac{1}{2} \div 2) \times \frac{1}{3}(2 - \frac{1}{2}) = \frac{1}{2}$ of $\frac{1}{4}$ of $\frac{1}{3}$ of $\frac{3}{2} = \frac{1}{16}$, *Ans.*

(8.) $\frac{1}{3} \times \frac{1}{2}$ of $(1 - \frac{1}{3}) \times \frac{1}{3}$ of $(2 - \frac{1}{3}) \times \frac{1}{5}$ of $(3 - \frac{1}{3}) \times \frac{1}{4}(4 - \frac{1}{3}) = \frac{1}{3} \times \frac{1}{2} \times \frac{1}{3} \times \frac{1}{5} \times \frac{1}{4}$ of $\frac{2}{3}$ of $\frac{5}{3}$ of $\frac{8}{3}$ of $\frac{11}{3} = \frac{22}{729}$, *Ans.*

(9.) $(2 + \frac{1}{5}) \div (3 + \frac{1}{7}) = \frac{11}{5} \div \frac{22}{7} = \frac{7}{10}$; $(2 - \frac{1}{3}) \times (4 - 3\frac{3}{7}) = \frac{5}{3} \times \frac{4}{7} = \frac{20}{21}$; $\frac{7}{10} \div \frac{20}{21} = \frac{147}{200}$, *Ans.*

(10.) $4\frac{1}{3} \times 4\frac{1}{3} \times 4\frac{1}{3} = \frac{13 \times 13 \times 13}{27} = \frac{2197}{27}$; $\frac{2197}{27} - 1 = \frac{2170}{27}$; $4\frac{1}{3} \times 4\frac{1}{3} - 1 = \frac{169}{9} - 1 = \frac{160}{9}$; $\frac{2170}{27} \div \frac{160}{9} = \frac{2170}{27} \times \frac{9}{160} = 4\frac{25}{48}$, *Ans.*

(11.) $\frac{2}{3}$ of $\frac{3}{4}$ of $\frac{7}{8} = \frac{7}{16}$; $\frac{1}{5} \times \frac{2}{3}$ of $\frac{15}{8} = \frac{1}{4}$; $\frac{7}{16} + \frac{1}{4} + \frac{1}{4} = \frac{7+8}{16} = \frac{15}{16}$, *Ans.*

(12.) $2\frac{1}{6} + \frac{7}{30} = \frac{72}{30}$; $\frac{3}{10} \div \frac{72}{30} = \frac{1}{8}$, the subtrahend; $\frac{1}{8}$ + remainder $\frac{25}{64} = \frac{33}{64}$, the minuend; this is $\frac{3}{5}$ of $\frac{10}{9}$, or, $\frac{2}{3}$ of some number; hence the number $= \frac{3}{2}$ of $\frac{33}{64} = \frac{99}{128}$, *Ans.*

(13.) $\frac{3}{4}$ of James's money $= \frac{3}{4}$ of $\frac{3}{5}$, or $\frac{9}{20}$, of Charles's money; hence $\frac{9}{20}$ of Charles's + \$33 $= \frac{20}{20}$ of it, which can be true only if $\frac{11}{20}$ of it = \$33; $\frac{1}{20}$ of it $= \frac{1}{11}$ of \$33, or \$3; $\frac{20}{20}$, or the whole = \$60; then $\frac{3}{5}$ of \$60 = \$36, James's, *Ans.*

(14.) They will meet in as many hours as the number of miles they both travel in 1 hour is contained times in 109 miles; 109 hr. $\div (7\frac{1}{2} + 8\frac{1}{4}) = 109 \times \frac{4}{63} = 6\frac{58}{63}$ hr., *Ans.*

Also, $6\frac{58}{63} \times 7\frac{1}{2}$ mi. $= \frac{436 \times 15}{126}$ mi. $= 51\frac{19}{21}$ mi., A; and $6\frac{58}{63} \times 8\frac{1}{4}$ mi. $= \frac{436 \times 33}{252} = 57\frac{2}{21}$ mi., B, *Ans.*

REMARK.—The sign $\times$ is often read "*times.*"

(15.) $\frac{5}{2}$ = product; $\frac{5}{9}$ of $\frac{3}{7}$ of $\frac{56}{15} = \frac{8}{9}$, one factor; the other factor $= \frac{5}{2} \div \frac{8}{9} = \frac{45}{16} = 2\frac{13}{16}$, *Ans.*

(16.) $14\frac{3}{4} \times 1\frac{3}{5} = \frac{59}{4} \times \frac{8}{5} = \frac{472}{20} = \frac{118}{5} = 23\frac{3}{5}$, *Ans.*

(17.) $29\frac{23}{35} - 14\frac{5}{8} = 15 + \frac{23}{35} - \frac{5}{8} = 15 + \frac{184-175}{280} = 15\frac{9}{280}$, *Ans.*

(18.) $\frac{5}{9} + \frac{1}{6} = \frac{13}{18}$; whole $- \frac{13}{18} = \frac{5}{18}$; $\frac{5}{18}$ of it $= \$60$; the whole $= \frac{18}{5}$ of $\$60 = \$216$, *Ans.*

(19.) $\frac{2}{3}$ of first $= \frac{3}{4}$ of second; $\frac{1}{3}$ of first $= \frac{1}{2}$ of $\frac{3}{4}$ of second; whole of first $= \frac{3}{2}$ of $\frac{3}{4}$ of second; hence $\frac{9}{8}$ of second + whole, or $\frac{17}{8}$ of it $= 51$; and second, therefore, $= \frac{8}{17}$ of $51, = 24$, *Ans.*
Also, $51 - 24 = 27$, *Ans.*

(20.) Rule.—*To find what part one number is of another, take the number which is the part for a numerator, and that of which it is the part for a denominator.*

Thus: $\dfrac{\frac{1}{2}}{\frac{2}{3}} = \frac{3}{4}$, *Ans.*

(21.) $\frac{7}{9} \times \frac{18}{5} \times \frac{14}{13} \times \frac{1}{7} = \frac{28}{65}$, *Ans.*
And, $\frac{9}{11} \times \frac{2}{3} \times \frac{55}{2} \times \frac{9}{4} \times \frac{17}{3} \times \frac{2}{11} = \frac{765}{22} = 34\frac{17}{22}$, *Ans.*

(22.) $\frac{7}{11} \times \frac{5}{2} \times \frac{3}{13} \times \frac{39}{2} = \frac{315}{44} = 7\frac{7}{44}$, *Ans.*
Also, $\frac{7}{8} \times \frac{5}{9} \times \frac{99}{7} \times \frac{11}{3} \times \frac{7}{2} \times \frac{9}{121} = \frac{105}{16} = 6\frac{9}{16}$, *Ans.*

(23.) Factor the denominators by finding their common divisors. (Art. 100, Rem.)

$\frac{3131}{71\times61} + \frac{1470}{41\times71} + \frac{1931}{41\times61} = \frac{355142}{41\times61\times71} = 2$; $2 \div (\frac{1}{7}$ of $\frac{5}{2})$ $= 2 \div \frac{5}{14} = 5\frac{3}{5}$, *Ans.*

(24.) $\frac{20}{20} - \frac{11}{20} = \frac{9}{20}$, younger son's share; difference of shares, $\frac{11}{20} - \frac{9}{20} = \frac{1}{10}$ of estate; estate, therefore, $= \$525 \times 10 = \$5250$, *Ans.*

(25.) $3\frac{7}{8} + 2\frac{1}{9} = 3\frac{63}{72} + 2\frac{8}{72} = 5\frac{71}{72}$, the *sum;* $3\frac{63}{72} - 2\frac{8}{72} = 1\frac{55}{72}$, the *difference; product* $= \frac{31}{8} \times \frac{19}{9} = 8\frac{13}{72}$; and $5\frac{71}{72} \div 1\frac{55}{72} = \frac{431}{72} \times \frac{72}{127} = 3\frac{50}{127}$, the *quotient.*

(26.) The ship is worth $\frac{1}{7}$ of the cargo; both $= \frac{8}{7}$ of cargo; $\frac{5}{16}$ of both $= \frac{5}{16} \times \frac{8}{7} = \frac{5}{14}$, *Ans.*

(27.) Factor denominators by finding G. C. D. of each two; then,

$\frac{757}{67 \times 41} + \frac{951}{67 \times 37} + \frac{517}{41 \times 37} = \frac{101639}{41 \times 37 \times 67} = 1$; $1000 \div 1 = 1000$, *Ans.*

(28.) $8128 = 1 \times 2 \times 2 \times 2 \times 2 \times 2 \times 2 \times 127$.

Divisors, 1 2 4 8 16 32 64 127 254 508 1016 2032 4064

Sum $= 8128$, a perfect number; $8128 \times 7 \div \frac{448}{3} = \frac{8128 \times 7 \times 3}{448} = 381$, *Ans*

(29.) $\frac{13}{2} \div \frac{117}{11} = \frac{11}{18}$, *Ans.*
$\frac{448+315}{504} - \frac{72+630}{504} = \frac{61}{504}$, *Ans.*

(30.) $4 \times \frac{3}{16} \times \frac{99}{7} \times \frac{11}{16} \times 5 \times \frac{3}{22} \times \frac{6}{5} \times \frac{4}{11} \times 6 = \frac{729}{56} = 13\frac{1}{56}$, *Ans.*

(31.) $\frac{175}{18} \times \frac{8}{7} \times \frac{9}{5} = 20$, *Ans.*

(32.) $\frac{5}{8}$ (full) $- \frac{9\frac{1}{2}}{63}$ (full) $= \frac{5}{8} - \frac{19}{126} = \frac{239}{504}$, *Ans.*

(33.) $3\frac{3}{4}$ mi. $\times 14\frac{3}{4} = \frac{15}{4} \times \frac{59}{4} = \frac{885}{16}$ mi.; $\frac{885}{16} \div 5\frac{1}{4} = \frac{885}{16} \times \frac{4}{21} = 10\frac{15}{28}$, the number of hours, *Ans.*

(34.) $(32\frac{3}{4} \times 17\frac{5}{8}) \div (17\frac{5}{8} - 4\frac{2}{3}) = \frac{131}{4} \times \frac{141}{8} \times \frac{24}{311} = \frac{55413}{1244}$ lb.; $\frac{55413}{1244}$ lb. $- \frac{131}{4}$ lb. $= 11\frac{247}{311}$ lb., *Ans.*

(35.) Whole $- \frac{2}{11} = \frac{9}{11}$, the first remainder; $\frac{9}{11}$ of the first amount $+ \$65 =$ amount next in hand; losing $\frac{3}{4}$ of this, he had remaining $\frac{1}{4}$, which was $\frac{9}{44}$ of the first amount $+ \$\frac{65}{4}$; with \$10 more, this would have been equal to the first amount; that is, $\frac{9}{44}$ of it $+ \$\frac{105}{4} =$ the whole; hence $\frac{35}{44}$ of it was $\$\frac{105}{4}$, and the whole was $\frac{44}{35}$ of $\$\frac{105}{4}$, $= \$33$, *Ans.*

DECIMAL FRACTIONS.

REDUCTION OF DECIMALS.

Art. 158.

(1.) $.25625 = 5)\frac{25625}{100000} = 5)\frac{5125}{20000} = 5)\frac{1025}{4000} = 5)\frac{205}{800} = \frac{41}{160}$, *Ans.*

(2.) $.15234375 = 5)\frac{15234375}{100000000} = 5)\frac{3046875}{20000000} =$
$5)\frac{609375}{4000000} — 5)\frac{121875}{800000} = 5\frac{24375}{160000} = 5)\frac{4875}{32000} =$
$5)\frac{975}{6400} = 5)\frac{195}{1280} = \frac{39}{256}$, *Ans.*

(3.) $.125 = 5)\frac{125}{1000} = 5)\frac{25}{200} = 5)\frac{5}{40} = \frac{1}{8}$. *Ans.* $2\frac{1}{8}$.

(4.) $.0175\emptyset = 5)\frac{175}{10000} = 5)\frac{35}{2000} = \frac{7}{400}$. *Ans.* $19\frac{7}{400}$.

(5.) $.00\frac{1}{5} = \frac{\frac{1}{5}}{100} = \frac{1}{5 \times 100} = \frac{1}{500}$. *Ans.* $16\frac{1}{500}$.

(6.) $.028\frac{4}{7} = \frac{28\frac{4}{7}}{1000} = \frac{200}{7000} = \frac{2}{70} = \frac{1}{35}$. *Ans.* $350\frac{1}{35}$.

(7.) $.666666\frac{2}{3} = \frac{666666\frac{2}{3}}{1000000}\frac{2000000}{3000000} = \frac{2}{3}$, *Ans.*

(8.) $.003125 = 5)\frac{3125}{1000000} = 5)\frac{625}{200000} = 5)\frac{125}{40000} =$
$5)\frac{25}{8000} = 5)\frac{5}{1600} = \frac{1}{320}$, *Ans.*

(9.) $.0\frac{5}{9} = \frac{\frac{5}{9}}{10} = \frac{5}{9 \times 10} = \frac{1}{18}$. *Ans.* $11\frac{1}{18}$.

(10.) $.390625 = 5)\frac{390625}{1000000} = 5)\frac{78125}{200000} = 5)\frac{15625}{40000} =$
$5)\frac{3125}{8000} = 5)\frac{625}{1600} = 5)\frac{125}{320} = \frac{25}{64}$, *Ans.*

(11.) $.1944\frac{4}{9} = \frac{19444\frac{4}{9}}{10000} = \frac{17500}{90000} = 5)\frac{175}{900} = 5)\frac{35}{180} = \frac{7}{36}$, *Ans.*

(12.) $.24\frac{4}{9} = \dfrac{24\frac{4}{9}}{100} = \frac{220}{900} = 2)\frac{22}{90} = \frac{11}{45}$, *Ans.*

(13.) $.33\frac{1}{3} = \dfrac{33\frac{1}{3}}{100} = \frac{100}{300} = \frac{1}{3}$, *Ans.*

(14.) $.66\frac{2}{3} = \dfrac{66\frac{2}{3}}{100} = \frac{200}{300} = \frac{2}{3}$, *Ans.*

(15.) $.25 = 5)\frac{25}{100} = 5)\frac{5}{20} = \frac{1}{4}$, *Ans.*

(16.) $.75 = 5)\frac{75}{100} = 5)\frac{15}{20} = \frac{3}{4}$, *Ans.*

Art. 159.

(1.) 4)3.00
.75, *Ans.*

(2.) 8)1.000
.125, *Ans.*

(3.) 20)1.00
.05, *Ans.*

(4.) 32)15.00000
.46875, *Ans.*

(5.) 16|00)9.0000|00
.005625, *Ans.*

(6.) 5)4.0
Ans. .8

(7.) 2|00)99.0|00
Ans. .495

(8.) 64)5.000000
Ans. .078125

(9.) 256)13.00000000
05078125, *Ans.*

(10.) 1024)1.0000000000
Ans. .0009765625

(11.) $\frac{1}{2} = \frac{1.0}{2} = .5$ *Ans.* 16.5

(12.) 16)3.0000
.1875; which, annexed to the 42, gives 42.1875, *Ans.*

(13.) $\frac{1}{4} = .25$; this, annexed to .015, gives .01525, *Ans.*

(14.) $\frac{3}{4} = .75$; this, annexed to 101.01, gives 101.0175, *Ans.*

(15.) 8|0)3.000|0
.0375; which, annexed to 75119, gives 75119.0375 *Ans.*

(16.) 32|0)1.00000|0

.003125 ; annexed to 2.00, gives 2.00003125, *Ans.*

ADDITION OF DECIMALS.

Art. 160.

(1.)
1.
.9475
Ans. 1.9475

(2.)
$1.33\frac{1}{3}$
$2.66\frac{2}{3}$
Ans. 4.00

(3.)
14.034
25.
.0000625
.0034
Ans. 39.0374625

(4.)
.083
21.01
2.5
94.5
Ans. 118.093

(5.)
$.166\frac{2}{3}$
.375
5.
$3.437\frac{1}{2}$
$.000\frac{7}{8}$
Ans. $8.980\frac{1}{24}$

(6.)
4.
.4
.04
Ans. 4.44

(7.)
$.111111\frac{1}{9}$
$.666666\frac{2}{3}$
$.222222\frac{2}{9}$
Ans. 1.000000

(8.)
$.1428\frac{4}{7}$
.0186
920.
$.0139\frac{3}{7}$
Ans. 920.1754

(9.)
$16.008777777\frac{7}{9}$
$.007466666\frac{2}{3}$
$.283333333\frac{1}{3}$
.000190422
Ans. $16.299768199\frac{7}{9}$

(10.)
.675
.000002
64.125
3.489107
.00089407
Ans. 68.29000307

(11.)
$4.067\frac{7}{8}$
$4.067\frac{7}{8}$
$4.067\frac{7}{8}$
$4.067\frac{7}{8}$
$.000\frac{1}{2}$
Ans. 16.272

(12.)
216.86301
48.1057
.029
1.3
1000.
Ans. 1266.29771

(13.)
35.
3.5
.35
.035
Ans. 38.885

(14.)
.010001
.00004
.96
.047060008
Ans. 1.017101008

Art. 161. SUBTRACTION OF DECIMALS.

(1.) 19.54
8.00717
Ans. 11.53283

(2.) 3000.
.003
Ans. 2999.997

(3.) 72.01
72.0001
Ans. $.00\overline{99}$

(4.) $1.169\frac{3}{7}$
$.93\frac{2}{35} = .930\frac{4}{7}$
Ans. $.238\frac{6}{7}$

(5.) 19.0
$8.999\frac{1}{9}$
Ans. $10.000\frac{8}{9}$

(6.) .4
$.04\frac{1}{3}$
Ans. $.35\frac{2}{3}$

(7.) .65007
$\frac{1}{2} = .5$
Ans. .15007

(8.) 2.75
1.8
Ans. .95

(9.) 1.684
1.
Ans. .684

(10.) $.000444\frac{4}{9}$
$.000000\frac{5}{6}$
Ans. $.000443\frac{11}{18}$

(11.) 4.9375
.015
4.9225

(12.) 10.
10.000
Ans. 0.

(13.) 3.701
2.45
Ans. 1.251

(14.) 1.875
$1\frac{7}{8} = 1.875$
Ans. 0.

(15.) $.0\frac{1}{18} = .00\frac{5}{9}$
$.00\frac{5}{9}$
Ans. 0.

(16.) 100.
$.64\frac{1}{6}$
Ans. $99.35\frac{5}{6}$

Art. 162. MULTIPLICATION OF DECIMALS.

(1.) 1
.1
Ans. .1

(2.) 16
$.03\frac{1}{3}$
48
$5\frac{1}{3}$
Ans. $.53\frac{1}{3}$

(3.) .01
.15
Ans. .0015

(4.) .080
80.
Ans. 6.4

(5.) 37.5
$82\frac{1}{2}$
750
3000
1875
3093.75, *Ans.*

(6.) 64.01
.32
12802
19203
Ans. 20.4832

(7.) 48000.
73.
144
336
Ans. 3504000.

(8.) 64.66$\frac{2}{3}$
18
51728
6466
12
1164.00, *Ans.*

(9.) .56$\frac{1}{4}$
.03$\frac{1}{16}$
168$\frac{3}{4}$
3$\frac{33}{64}$
.0172$\frac{17}{64}$
Ans.

(10.) 738.
120.4
2952
1476
738
88855.2, *Ans.*

(11.) 1.006
.0001
.0001006
Ans.

(12.) .193
34.
772
579
Ans. 6.562

(13.) 2.7
.42
54
108
Ans. 1.134

(14.) 43.7004
.008
Ans. .3496032

(15.) 21.0375
4.44$\frac{1}{9}$
841500
841500
841500
93500
93.5, *Ans.*

(16.) 9300.701
251
9300701
46503505
18601402
2334475.951, *Ans.*

(17.) 430.0126
4000.
1720050.4$\not{0}\not{0}\not{0}$
Ans.

(18.) .059
.059
531
295
.003481
.059
31329
17405
.000205379, *Ans.*

(19.) 4 2.
4.2
8 4
168
176.4, *Ans.*

(20.) .02$\frac{1}{4}$
600.
1200
150
13.5, *Ans.*

(21.) 7100.0
.0000001$\frac{1}{8}$
Ans. .0008875

(22.) 26000000.
.000026
156
52
Ans. 676.

(23.) 27.00
6.0
Ans. 162.

```
(24.)  6.3029
        .03275
       ---------
        315145
     189087
      1701783
     -----------
     .206419975, Ans.
```

```
(25.)  135.027
         1.00327
      -----------
       405081
         3645729
     135027
     -------------
     135.46853829, Ans.
```

Art. 163.

```
         7519
(1.)  27.6530
        9.157
      --------
      248877|0
        2765|3
        1382|7
         193|6
      --------
     253.218 6
     253.219,
        Ans.
```

```
         951413
(2.)   4320710
        314159
      ---------
      1296213|0
        43207|1
        17282|8
          432|1
          217|0
           38|9
      ---------
     135.7389 9
Ans. 135.7390
```

```
          23461
(3.)   3.62741
        1.6432
      --------
       36274|1
       21764|5
        1451|0
         108|8
           7|3
      --------
      5.9605 7
Ans.  5.9606
```

```
        5 784
(4.)   9.012
       48.75
      -------
       3604|8
        721|0
         63|0
          4|5
      -------
Ans.  439.3 3
```

```
        957010
(5.)  4.804136
       .010759
      --------
       48041|4
        3362|9
         240|2
          43|2
      --------
       51687 7
Ans.  .051688
```

```
        5454 59762
(6.)   814.0737134
        26.7954545
      -------------
       16281474|3
        4884442|3
         569851|6
          73266|6
           4070|4
            325|6
             40|7
              3|3
               |4
      -------------
Ans.  21813.475 2
```

(7.)

```
     4491 8521
     702.6100
    1.2581944
    ---------
    702610|0
    140522|0
     35130|5
      5620|9
        70|3
        63|2
         2|8
          |3
    ---------
Ans. 884.020 ø
```

(8.)

```
     4444 240
     849.9375
     .0424444
     --------
     33997|5
      1699|9
       340|0
        34|0
         3|4
          |3
     --------
Ans. 36.075 1
```

(9.)

```
      27 131
     880.695
      131.72
     -------
     88069|5
     26420|9
       880|7
       616|5
        17|6
     -------
Ans. 116005.2
```

(10.)

```
     709400
     .02538 1
     .004907
     -------
        10|1
         2|3
     -------
Ans. .00012 4
```

(11.)

```
      73530
     64.010820
        .03537
     ---------
     1920324|6
      320054|1
       19203|2
        4480|7
     ---------
     2.264062 6
Ans. 2.264063
```

(12.)

```
     3843 2
     1380.375
       .23483
     --------
       27607|5
        4141|1
         552|1
         110|4
           4|1
            |4
     --------
       324.15 6
Ans. 324.16
```

DIVISION OF DECIMALS.

Art. 165.

(1.) $\frac{63.00000}{4000} = .01575$, *Ans.*

(2.) $\frac{3.1500}{375} = .0084$, *Ans.*

(3.) $\frac{1.008}{18} = .056$, *Ans.*

(4.) $\frac{4096.000}{.032} = 128000$, *Ans.*

(5.) $\frac{9.7000}{97000} = .0001$, *Ans.*

(6.) $\frac{.9}{.00075} = \frac{3600}{3} = 1200$, *Ans.*

(7.) $\frac{13.0000}{78.12\frac{1}{2}} = \frac{13\times8}{625} = \frac{104.0000}{5\times5\times5\times5} = .1664$, *Ans.*

(8.) $\frac{12.9000000}{8.256} = 1.5625$, *Ans.*

(9.)
$\frac{81.20960}{1.28} = 63.445$, *Ans.*

(10.)
$\frac{1.00}{100} = .01$, *Ans.*

(11.)
$\frac{10.10000}{17} = .59412-$, *Ans.*

(12.)
$\frac{.00100}{100} = .00001$, *Ans.*

(13.)
$\frac{12755.00000}{81632} = .15625$, *Ans.*

(14.)
$\frac{2401.0000}{21.4375} = 112$, *Ans.*

(15.)
$\frac{21.13212}{.916} = 23.07$, *Ans.*

(16.)
$\frac{36.7267200}{.5025} = 73.088$, *Ans.*

(17.) $\frac{2483.25 \times 4 \times 4 \times 4}{5.15625 \times 4 \times 4 \times 4} = \frac{158928}{330} = 481.6$, *Ans.*

REMARK.—Such examples as the 7th and the 17th should be performed both ways, that the pupil may compare the two processes—the one an easy reduction, and the other a tedious division.

(18.) $\frac{142.0281000}{9.2376} = 15.375$, *Ans.*

(19.) $\frac{.08\frac{1}{3}}{.12\frac{1}{2}} \times \frac{600}{600} = \frac{50}{75} = .66\frac{2}{3} = \frac{2}{3}$, *Ans.*

(20.) $\frac{.0001}{.01} = .01$, *Ans.*

(21.) $\frac{95.300000000}{.264} = 360.984848+$, *Ans.*

(22.) $\frac{1000}{.001} = 1000 \div \frac{1}{1000} = 1000 \times \frac{1000}{1} = 1000000$, *Ans.*

(23.) $10 \div .1 = 10 \div \frac{1}{10} = 10 \times 10 = 100$, *Ans.*

(24.) $\frac{.000001}{.01} = .0001$, *Ans.*

(25.) $\frac{.00001000}{1000} = .00000001$, *Ans.*

(26.) $\frac{16.275000000}{.41664} = 39.0625$, *Ans.*

(27.) $\frac{1}{10000000} \div \frac{1}{100} = \frac{1}{10000000} \times \frac{100}{1} = .00001$, *Ans.*

ABBREVIATED DIVISION.

Art. 166.

(1.)

```
.98)1000.0000(1020.408
    98        1020.41,
    --
     200          Ans.
     196
     ---
      400
      392
      ---
       80
       72
       --
```

(2.)

```
.995)6215.75000(6246.985, Ans.
     5970
     ----
      2457     9800
      1990     8955
      ----     ----
       4675     8450
       3980     7960
       ----     ----
        6950     490
        5970     498—
        ----     ---
         9800
```

(3.)

```
.993)28012.0000(28209.466
      1986        28209.47, Ans.
     -----
       8152
       7944
       ----
        2080
        1986
        ----
         9400
         8937
         ----
          4630
          3972
          ----
           658
           596
           ---
            62
```

(4.)

```
.9975)52546.35000(52678.045,
       49875               Ans.
       -----
        26713
        19950
        -----
         67635
         59850
         -----
          77850
          69825
          -----
           80250
           79800
           -----
             4500
             3990
             ----
              510
              499
              ---
               11
```

(5.)

```
.9875)4840.0000(4901.266
      39500     4901.27, Ans.
      -----
       89000     2625
       88875     1975
       -----     ----
        12500     650
         9875     593
         ----     ---
         2625      57
```

(6.) $1.4142136)2.0000000\dot{0}(1.4142135$, *Ans.*

14142136	
58578640	
56568544	
2010096	1913
1414214	1414
595882	499
565685	424
30197	75
28284	71
1913	4

(7.) $3.14159265)9.86960440\dot{1}(3.14159265$, *Ans.*

942477795	
444826451	291091
314159265	282743
130667186	8348
125663706	6283
5003480	2065
3141593	1885
1861887	180
1570796	157
291091	23

CIRCULATING DECIMALS.

Art. 173. CASE II.

(1.)
$.\dot{3} = \frac{3}{9} = \frac{1}{3}$, *Ans.*

(2.)
$.0\dot{5} = \frac{5-0}{90} = \frac{1}{18}$, *Ans.*

(3.)
$.\dot{1}2\dot{3} = \frac{123}{999} = \frac{41}{333}$, *Ans.*

(4.)
$2.\dot{6}\dot{3} = 2\frac{63}{99} = 2\frac{7}{11}$, *Ans.*

(5.) $.3\dot{1} = \frac{31-3}{90} = \frac{14}{45}$, *Ans.*

(6.) $.0\dot{2}1\dot{6} = \frac{216-0}{9990} = \frac{4}{185}$, *Ans.*

(7.) $\dot{4}8.\dot{1} = 48.\dot{1}4\dot{8} = 48\frac{148}{999} = 48\frac{4}{27}$, *Ans.*

(8.) $\dot{1}.00\dot{1} = 1.\dot{0}01\dot{1} = 1\frac{11}{9999} = 1\frac{1}{909}$, *Ans.*

(9.) $.13\dot{8} = \frac{138-13}{900} = \frac{125}{900} = \frac{5}{36}$, *Ans.*

(10.) $.208\dot{3} = \frac{2083-208}{9000} = \frac{1875}{9000} = \frac{5}{24}$, *Ans.*

(11.) $\dot{8}5.714\dot{2} = 85.\dot{7}1428\dot{5} = 85\frac{714285}{999999} = 85\frac{5}{7}$, *Ans.*

(12.) $.\dot{0}6349\dot{2} = \frac{63492}{999999} = \frac{4}{63}$, *Ans.*

(13.) $.4\dot{4}7619\dot{0} = \frac{4476190-4}{9999990} = \frac{47}{105}$, *Ans.*

(14.) $.0902\dot{7} = \frac{9027-902}{90000} = \frac{13}{144}$, *Ans.*

ADDITION OF CIRCULATES.

Art. 174.

(1.)	(2.)	(3.)	(4.)
$.45\dot{3}$	$3.04\dot{4}\dot{4}$	$.25\dot{2}\dot{5}$	$1.03\dot{1}0310\dot{3}$
$.06\dot{8}$	$6.45\dot{6}\dot{6}$	$.10\dot{4}\dot{4}$	$.25\dot{7}7777\dot{7}$
$.32\dot{7}$	$23.38\dot{3}\dot{8}$	$.61\dot{6}\dot{1}$	$5.04\dot{0}4040\dot{4}$
$.94\dot{6}$	$.24\dot{8}\dot{4}$	$.56\dot{3}\dot{5}$	$28.0445245\dot{2}$
Ans. $1.79\dot{6}$	*Ans.* $33.13\dot{3}\dot{4}$	$1.53\dot{6}\dot{6}$	$34.37\dot{3}7373\dot{7}$
		Ans. $1.53\dot{6}$	*Ans.* $34.\dot{3}\dot{7}$

(5.)	(6.)	(7.)
$.666\dot{6}$	$9.21\dot{1}0710\dot{7}$	$.20\dot{4}\dot{5}$
$.138\dot{8}$	$.65\dot{6}5656\dot{5}$	$.09\dot{0}\dot{9}$
$.055\dot{5}$	$5.00\dot{4}4444\dot{4}$	$.25\dot{0}\dot{0}$
$.097\dot{2}$	$3.56\dot{2}2622\dot{6}$	$.54\dot{5}\dot{4}$
$.041\dot{6}$	$18.43\dot{4}3434\dot{3}$	*Ans.* $.\dot{5}\dot{4}$
Ans. $1.000\dot{0}$	*Ans.* $18.\dot{4}\dot{3}$	

(8.) $5.\dot{0}7707\dot{7}$
$.\dot{2}4242\dot{4}$
$7.\dot{1}2494\dot{3}$
$12.\dot{4}4444\dot{4}$
Ans. $12.\dot{4}$

(9.) $3.48\dot{8}44\dot{8}$
$1.63\dot{7}37\dot{3}$
$130.81\dot{1}11\dot{1}$
$.06\dot{6}66\dot{6}$
$136.00\dot{3}60\dot{0}$
Ans. $1\dot{3}6.0\dot{0}$

SUBTRACTION OF CIRCULATES.

Art. 175.

(1.) $.2\dot{6}6\dot{6}$
$.0\dot{0}7\dot{4}$
$.2\dot{5}9\dot{2}$
Ans. $.\dot{2}5\dot{9}$

(2.) $15.35\dot{4}6\dot{5}$
$9.09\dot{9}0\dot{9}$
$6.25\dot{5}5\dot{5}$
Ans. $6.2\dot{5}$

(3.) $18.23\dot{6}7\dot{3}$
$4.51\dot{4}5\dot{1}$
$13.72\dot{2}2\dot{2}$
Ans. $13.7\dot{2}$

(4.) $100.73\dot{0}\dot{0}$
$37.01\dot{2}\dot{8}$
$63.71\dot{7}\dot{1}$
Ans. $63.\dot{7}\dot{1}$

(5.) $10.0\dot{5}6356\dot{3}$
$8.2\dot{7}2727\dot{2}$
Ans. $1.7\dot{8}3629\dot{0}$

(6.) $199.6\dot{4}2857\dot{1}$
$190.4\dot{7}6190\dot{4}$
$9.1\dot{6}6666\dot{6}$
Ans. $9.1\dot{6}$

(7.) $104.\dot{1}4141\dot{4}$
$13.\dot{6}3763\dot{7}$
Ans. $90.\dot{5}0377\dot{6}$

MULTIPLICATION OF CIRCULATES.

Art. 176.

(1.) $4.7\dot{3}\dot{5}$
7.349
$.0426\dot{1}\dot{8}$
$.1894\dot{1}\dot{4}$
$1.4206\dot{0}\dot{6}$
$33.1474\dot{7}\dot{4}$
$34.8001\dot{1}\dot{3}$, *Ans.*

(2.) $.0\dot{7}06\dot{7}$
$.9\dot{4}3\dot{2} = .9\frac{16}{37}$
$.06\dot{3}60\dot{9}$
$.00\dot{3}05\dot{6}$
$.06\dot{6}66\dot{5}$, *Ans.*

(3.) $714.3\dot{2}$
$3.45\frac{2}{3}$
$35.716\dot{1}1\dot{1}$
$285.728\dot{8}8\dot{8}$
$2142.966\dot{6}6\dot{6}$
$4.762\dot{1}4\dot{8}$
$2469.173\dot{8}1\dot{4}$, *Ans.*

(4.) $16.\dot{2}0\dot{4}$
$32\frac{25}{33}$
$32.\dot{4}08408\dot{8}$
$486.\dot{1}26126\dot{6}$
$12.\dot{2}75912\dot{2}$
$530.\dot{8}10446\dot{6}$, *Ans.*

(5.) $19.0\dot{7}\dot{2}$
$.208\frac{1}{3}$
$.1525\dot{8}\dot{1}$
$3.8145\dot{4}\dot{5}$
$.0063\dot{5}\dot{7}$
$3.9734\dot{8}\dot{4}$
$3.973\dot{4}\dot{8}$, *Ans.*

(6.) $3.\dot{7}54\dot{3}$
4.7157
$.0026\dot{2}80\dot{6}$
$.0187\dot{7}18\dot{7}$
$.0375\dot{4}37\dot{5}$
$2.6280\dot{6}28\dot{0}$
$15.0175\dot{0}17\dot{5}$
$17.7045\dot{0}82\dot{5}$
$17.704\dot{5}08\dot{2}$, *Ans.*

(7.) $1.\dot{2}5678\dot{4}$
$6.420\frac{9}{11}$
$.025\dot{1}3568\dot{5}$
$.502\dot{7}1370\dot{2}$
$7.540\dot{7}0554\dot{0}$
$1\dot{0}2827\dot{8}$
$8.069\dot{5}8320\dot{6}$, *Ans.*

DIVISION OF CIRCULATES.

Art. 177.

(1.) $.\dot{1}\dot{1})\underline{.\dot{7}\dot{5}}$
Ans. $6.\dot{8}\dot{1}$

(2.) $17.)51.49\dot{1}(3.02\dot{8}$, *Ans.*
51
49
34
151
136
15

(3.) $.94)681.55\dot{9}887\dot{9}(7.25\dot{0}637\dot{1}$, *Ans.*
658
235
188
475
470
598
564
348
282
667
667
658
99
94
5

(4.) $6.7\dot{5}45454\dot{4})90.5\dot{2}03749\dot{9}$

67 905

67545387) 905202844(13.$\dot{4}0\dot{1}$, *Ans.*

67545387

229748974

202636161

271128130

270181548

94658200

67545387

27112813

(5.) $.\dot{2}45245245\dot{5})11.\dot{0}68735402\dot{2}$

11

245245245) 11068735391(45.1$\dot{3}$, *Ans.*

980980980

1258925591

1226226225

326993660

245245245

817484150

735735735

81748415

(6.) $6.2\dot{1}72172172\dot{2})9.5\dot{3}30663997\dot{7}$

62 95

6217217211|0)9533066390|2(1.5$\dot{3}$, *Ans.*

6217217211

33158491792

31086086055

20724057370

18651651633

2072405737

(7.) $7.684\dot{4}4444444\dot{4})3.500\dot{6}91358024\dot{4}$

$$\begin{array}{r}
7684 \qquad\qquad 3500 \\
\hline
768444443676|0\;)\;350069135452|4(.4\dot{5},\ \textit{Ans.} \\
3073777774704 \\
\hline
4269135798200 \\
3842222218380 \\
\hline
426913579820
\end{array}$$

REDUCTION OF COMPOUND NUMBERS.

Art. 220.

(1.) $18.22 \times 4.76 \times 16 = 1387.6352$ sq. rd. *Ans.*

REMARK.—Strictly, the steps of a solution require some expressions different from the mere numerical indications. The solution just given is like the first model on page 156, New Higher Arith., and is quite convenient. Taking the three as *abstract* numbers, the statement is not accurate. Usage may allow this for the convenience of the operator, but pupils should be often drilled upon the solutions, thus:

18.22 ch. $\times$ 4.76 ch. $= 86.7272$ sq. ch.; and
86.7272 sq. rd. $\times 16 = 1387.6352$ sq. rd., *Ans.*

(2.) $16.02 \div 80 = .20025$ mi., *Ans.*

(3.) $750 \times 2.8375 = 21.28\frac{1}{8}$, bu., *Ans.*

(4.) 35.781 sq. yd. $= 35.781 \times 9 \times 144 = 46372.176$ sq. in., *Ans.*

(5.) $10240 \div 16 = 640$ sq. ch., *Ans.* (Art. 198.)

(6.) $40 \times 7\frac{1}{2} \times 2\frac{2}{3} \div 24\frac{3}{4} = 40 \times \frac{15}{2} \times \frac{8}{3} \times \frac{4}{99} = 32\frac{32}{99}$ P., *Ans.*

(7.) One carat $= 3.168$ gr.

$\frac{1680 \times 3.168}{480} = 21 \times .528 = 11.088$ oz., *Ans.*

(8.) 75 pwt. $= 75 \times 24 = 1800$ gr. tr. $= 1800 \times \frac{1}{60} =$ 30 ʒ, *Ans.*

(9.) $\frac{4}{7}$ gr. $= \frac{4}{7} \times \frac{1}{20} \times \frac{1}{3} \times \frac{1}{8} = \frac{1}{840}$ ℥, *Ans.*

(10.) $18\frac{3}{4}$ ʒ $= \frac{75}{4} \times 60 = 1125$ gr.; 1125 gr. $\div \frac{7000}{16} = \frac{18}{7} = 2\frac{4}{7}$ oz., *Ans.*

(11.) 96 oz. av. $\times \frac{1}{16} \times \frac{175}{144} \times 12 = 87\frac{1}{2}$ oz. tr., *Ans.*

Or, at length, thus:

1 oz. tr. $= \frac{1}{12}$ of $\frac{144}{175}$ lb. av. $= \frac{12}{175}$ of 16 oz. av. $= \frac{2}{175}$ of 96 oz. av. Hence, 96 oz. av. $= \frac{175}{2}$, or $87\frac{1}{2}$ oz. tr., *Ans.*

(12.) $3 \times \frac{9}{4} \times \frac{3}{2} \times \frac{1728}{231} = \frac{216 \times 27}{77} = 75\frac{57}{77}$ gal., *Ans.*

(13.) $\frac{9.3 \times 3.625 \times 2.25 \times 1728}{2150.4} = \frac{21845.7}{358.4} = 60.9$, or 61 bu., nearly, *Ans.*

(14.) $75 \div .2759 = 271.837+$ s., *Ans.*

(15.) $2\frac{1}{4}$ yr. $= 2\frac{1}{4} \times 31536000$ sec. $= 70956000$ sec., *Ans.*

(16.) 1 wk. $= 168$ hr.; hence $\frac{49}{168}$, or $\frac{7}{24}$, *Ans.*

(17.) 90.12 kl. $= 90.12$ l. $\times 1000 = 90120$ l., *Ans.*

(18.) $25'' = \frac{25}{3600}$ degrees; therefore $.0069\frac{4}{9}$, or $.0069\dot{4}^\circ$, *Ans.*

(19.) 1 sq. yd. $= 1296$ sq. in.; hence $\frac{192}{1296}$, or $\frac{4}{27}$ sq. yd., *Ans.*

(20.) $6\frac{2}{3}$ cu. yd. $= \frac{20}{3}$ cu. in. $\times 27 \times 1728 = 311040$ cu. in., *Ans.*

(21.) \$117.14 $= 117.14$ mills $\times 1000 = 117140$ mills, *Ans.*

(22.) 6.19 ct $=$ \$6.19 $\div 100 =$ \$.0619, *Ans.*

(23.) 1600 mills $= .001$ of \$1600 $=$ \$1.60, *Ans.*

(24.) \$$5\frac{3}{8}$ $=$ \$5.375 $= 5.375$ mills $\times 1000 = 5375$ mills, *Ans.*

(25.) 12 lb. av. $= \frac{175}{144}$ of 12 ℔. tr. $= 14\frac{7}{12}$ ℔., *Ans.*

(26.) $6.45 \times 100 = 645$, the number of kilograms; and $645 \times 1000 = 645000$ g., *Ans.* (See Art. 215.)

(27.) 1 oz. tr. $= \frac{1}{12}$ of 5760, or 480 gr.; $\frac{.216}{480} = .00045$ oz. tr., *Ans.*

(28.) 47.3084 sq. rd. $\times 640 \times 160 = 4844380.16$ sq. rd., *Ans.*

(29.) $4\frac{1}{2}$ ℈ $= \frac{1}{3}$ of $\frac{1}{8}$ of $\frac{1}{12}$ of $\frac{9}{2}$ ℔. $= \frac{1}{64}$ ℔., *Ans.*

(30.) $7\frac{1}{9}$ oz. $= \frac{1}{16}$ of $\frac{1}{100}$ of $\frac{64}{9}$ cwt. $= \frac{1}{225}$ cwt., *Ans.*

(31.) 99 yd. $= \frac{1}{1760}$ of 99 mi. $= \frac{9}{160}$ mi., *Ans.*

(32.) $\frac{16.02 \times 24.5}{160}$ acres $= 2.4530625$ acres, *Ans.*

(33.) $\frac{25}{4}$ of $\frac{5}{2}$ of $3 \times \frac{1}{27}$ cu. yd. $= 1\frac{53}{72}$ cu. yd., *Ans.*

(34.) 169 ars $= 100 \times 169$ $m^2 = 16900$ m^2, *Ans.*

(35.) $2\frac{1}{2}$ f℥ $= \frac{5}{2} \times 8 \times 60$ ♏ $= 1200$ ♏, *Ans.*

(36.) $\frac{6}{7}$ pure $= \frac{6}{7}$ of 24 carats fine $= 20\frac{4}{7}$ carats, *Ans.*

(37.) $18\frac{3}{4}$ carat gold is $\frac{18\frac{3}{4}}{24}$, or $\frac{25}{32}$ pure; and $1 - \frac{25}{32} = \frac{7}{32}$ alloy, *Ans.*

(38.) Width by breadth $= 6.15 \times 5.03$ $m^2 = 30.9345$ m^2., *Ans.*

(39.) $120 \times \frac{13}{2} \times \frac{35}{4} \times \frac{1}{128} = 53\frac{41}{128}$ C., *Ans.*

(40.) $2 \times 21\frac{1}{2} \times 13$ sq. ft. $+ 2 \times 16\frac{1}{2} \times 13$ sq. ft. $=$ $(43 + 33) \times 13$, or 988 sq. ft., *Ans.*

(41.) 18 cwt. 3 qr. 15 lb. 12 oz. $= \frac{30252}{32000}$, or $.945\frac{3}{8}$ T.; $\$48.20 \times 27.945\frac{3}{8} = \$1346.97$—, *Ans.*

(42.) $16.15 \times 1.22 \times 1.68$ times $\$2.30 = \$76.13+$, *Ans.*

(43.) $\frac{178 \times 84}{160} \times \$25.60 = \$2325.12$, *Ans.*

(44.) 160 sq. rd., 85 sq. in. $= 6272725$ sq. in.; $6272725 \div 25 = 250909$ stones, *Ans.*

(45.) Area $= 20 \times 15$, or 300 m^2.; 300 m^2. $- (4 + 9)$ m^2. $= 287$ m^2.; $\$.51 \times 287 = \$146.37$, *Ans.*

(46.) Solidity $= \frac{63\frac{1}{4} \times 38 \times 20}{1728 \times 27}$ cu. yd.

Weight $= \frac{63\frac{1}{4} \times 38 \times 20 \times 19128 \times 16}{1728 \times 27}$ oz. $= 315323.374$ oz., $= 9$ T. 17 cwt. 7 lb. 11.37 oz.

(47.) 4 pwt. 9 gr. $= 105$ gr. $= \frac{7}{32}$, or $.218\frac{3}{4}$ oz. Weight $= 3.218\frac{3}{4}$ oz.; cost $= \$2.20 \times 3.218\frac{3}{4} = \$7.08125$; weight $= 13$ pwt., and cost $= \$1\frac{1}{4} \times 13 = \$16.25$. Sum of costs $= \$23.331$, *Ans.*

REMARK.—In cases like this it will be a useful review exercise to give the solutions the bill form, as in Art. 66, New Higher Arith.

(48.) Capacity $= \frac{21}{2} \times \frac{7}{2} \times \frac{3}{2} \times 1728$ cu. in., $\frac{1}{2}$ of which in bu. $= \frac{21 \times 7 \times 3 \times 1728}{16 \times 2150.4} = 22.148$ bu.

22.148 bu. $= 22$ bu. 4 qt. 1.5— pt., *Ans.*

(49.) 160 lb. av. $= \frac{175}{144}$ of 160 ℔. tr. $= 194\frac{4}{9}$ ℔. $= 194$ ℔., 5 oz., 6 pwt., 16 gr., *Ans.*

(50.) 50 mi. $= 264000$ ft.

$264000 \div 13\frac{1}{2} = 19555.55+$, revolutions.

$264000 \div 18\frac{1}{3} = 14400.$ 〃

Ans. 5155. revolutions.

(51.) 5 ℔ 10 ℥ = 70 ℥, and, at \$2.20 per ℥, cost \$154; 1 gr. costing $\frac{10}{9}$ ct., 70 ℥, or 3360 gr., cost \$373.33$\frac{1}{3}$; diff. = \$219.33$\frac{1}{3}$, *Ans.*

(52.) Distance = 275 × 5280 feet; and 5280 × 275 ÷ 2.75 = 528000 steps, *Ans.*

(53.) 65350 sq. mi. = $\frac{65350 \times 640}{2.471}$ = 16925940.91+ Ha., *Ans.*

(54.) $\frac{36 \times 24 \times 14 \times 1728}{231}$ = 90484.36+ gal., *Ans.*

(55.) 60 × 15 × 2 × 8 = 14400 shingles, *Ans.*

(56.) In the two years there are 731 days. Since the leap year, 1884, commences neither on Saturday nor on Sunday, it has only 52 Sundays; 1885, commencing on Thursday, has only 52 Sundays; hence, deducting 104 days, we have 627 days, each with a gain of 70 min., in all 43890 min. = 731 hr. 30 min., *Ans.*

COMPOUND ADDITION.

Art. 221.

(1.)

mi.	rd.	ft.	in.
0	91	7	.857$\frac{1}{7}$
	146	5	6.
10	14	7	6.
	209	9	10.8
	37	16	2.25
1	0	12	8.726
12	180	8$\frac{1}{2}$	10.633$\frac{1}{7}$
		$\frac{1}{2}$ =	6.

12 mi. 180 rd. 9 ft. 4.633$\frac{1}{7}$ in., *Ans.*

(2.)

yd.	ft.	in.
6	0	6.84
2	2	9.75
	1	4.54
10	2	4.512
	2	3
	1	10.
		.875
21	2	3.517

Ans.

(3.)

yd.	qr.	na.	in.
3	2	3	1.5
	1	2	.9
6	0	1	2.175
1	2	2	.18
		2	1.5
			1.25
12	0	1	0.755, *Ans.*

(4.)

1.
.06
.957
.14
1414.2
734.8
82.
110.6
1407124.
.018
1409467.775, *Ans.*

(5.)

fr.	d.	c.
9	7	6
	6	8
1	3	7
2	6	5
3	0	4
17½ fr., *Ans.*		

(6.)

sq. yd.	ft.	sq. in.
15	5	87.
16	4	72.
10	7	31.68
	4	121.6
	3	13.5
43	7	37.78, *Ans.*

(7.)

A.	sq. rd.
101	98.35
66	74.5
20	0.
12	113.
5	13.33⅓
205	139.18⅓, *Ans.*

(8.)

cu. yd.	cu. ft.	cu. in.
23	14	1216.
41	6	642.132
9	25	112.32
	12	1036.8
75	4	1279.252, *Ans.*

(9.)

cu. ft.	cu. in.
106	1152
	648
	1000
107	1072
	Ans.

(10.)

lb.	oz.	pwt.	gr.
2	6	13	8.
1	9	0	0.
		12	16.32
	11	13	19.2
	4	10	0.
		18	18.
			$13.33\frac{1}{3}$
5	9	9	$2.85\frac{1}{3}$, *Ans.*

(11.)

lb.	℥	ʒ	℈	gr.
0	8	0	0	14.6
	4	1	1	6.4
		7	1	5
		2	2	18
	1			12
				4
1	2	4	1, *Ans.*	

(12.)

T.	cwt.	qr.	lb.
0	6	1	0
	9	1	22
		3	$1\frac{1}{2}$
4	8	3	$1\frac{2}{5}$
		3	6
		2	$8\frac{1}{3}$
5	6	2	$14\frac{7}{30}$, *Ans.*

(13.) *a.*

$4\frac{4}{5}$ oz.
$\frac{5}{9}$ "
$5\frac{16}{45}$ oz.

(13.) *b.*

4 oz.	$12\frac{4}{5}$ dr.
	$8\frac{8}{9}$
5 oz.	$5\frac{31}{45}$ dr., *Ans.*

REMARK.—The teacher may present both of these solutions, and remind the pupil of the difference between a *concrete* addition and a *compound* addition. (Art. 178, 5.)

(14.)

gal.	qt.	pt.
6	3	$\frac{2}{3}$
2	1	$\frac{83}{100}$
1	2	$\frac{1}{2}$
	2	$\frac{4}{5}$
		$\frac{4}{9}$
		$\frac{7}{8}$
11	2	.11$\frac{11}{18}$, *Ans.*

(15.)

gal.	qt.	pt.
4	0	.75
10	3	1.5
8	0	.66$\frac{2}{3}$
5	2	1.12
	2	.6
		1.27
		.15
29	2	.05$\frac{2}{3}$, *Ans.*

(16.)

bu.	pk.	qt.	pt.
1	0	4	0.
	2	1	.90$\frac{10}{11}$
	3	5	1.25
9	3	2	.48
		7	1.16
		3	.66$\frac{2}{3}$
12	3	0	.46$\frac{19}{33}$, *Ans.*

(17.)

pk.	qt.	pt.
1	4	0
	3	$\frac{2}{5}$
		1$\frac{5}{9}$
		$\frac{2}{3}$
2	0	$\frac{28}{45}$, *Ans.*

(18.)

f℥	fʒ	♏
6	2	25
2	4	0
	7	42
1	2	40
3	6	51
14	7	38, *Ans.*

(19.)

d.	hr.	min.	sec.
3	12	0	0
	12	30	0
			30$\frac{1}{2}$
4	0	30	30$\frac{1}{2}$, *Ans.*

(20.)

yr.	d.	hr.	min.	sec.
3	94	21	36	0
	118	5	42	$37\frac{1}{2}$
	63	9	36	0
		7	48	0
1	62	19	24	48
		9	7	30
4	340	1	14	$55\frac{1}{2}$, *Ans.*

(21.)

°	′	″
27	14	55.24
9	0	18.25
1	15	20.
116	44	23.8
154	14	57.29
		Ans.

(22.)

ct.	m.
50	0
	$2\frac{1}{2}$
	$\frac{1}{8}$
50	$2\frac{5}{8}$, *Ans.*

(23.)

$	ct.	m.
3	0	7
5	20	0
100	2	6
19	0	$1\frac{1}{4}$
127	23	$4\frac{1}{4}$, *Ans.*

(24.)

£	s.	d.
21	6	3.5
5	17	9.
9	1	8.4
	16	7.25
	8	4.
37	10	8.15. *Ans.*

(25.)

	sq. rd.	sq. yd.	sq. ft.	sq. in.
$\frac{2}{3}$ A. =	106	20	1	72
$\frac{3}{4}$ sq. rd. =		22	6	27
$\frac{4}{5}$ sq. ft. =				$115\frac{1}{5}$
	107	$11\frac{3}{4}$	8	$70\frac{1}{5}$
		$\frac{1}{4}$ =	(2	36)
	107	12	6	$34\frac{1}{5}$ *Ans.*

COMPOUND SUBTRACTION.

Art. 222.

(1.)

mi.	rd.
	144.86
$\frac{2}{5}$ =	128.
Ans.	16.86

(2.)

yd.	qr.	na.	in.
4	2	1	$1\frac{3}{4}$
1	1	1	$1\frac{7}{20}$
3	1	0	$\frac{2}{5}$, *Ans.*

(3.)

sq. rd	sq. yd.	sq. ft.	sq. in.
5	16	6	96
2	24	0	91
2	$22\frac{1}{4}$	6	5
2	22	8	41, *Ans.*

(4.)

sq. m.	A.	sq. rd.
1	195	32
	384	43.92
	450	148.08, *Ans.*

(5.)

cu. yd.	cu.ft.	cu. in.
20	4	1000
13	25	1204.9
6	5	1523.1, *Ans*

(6.)

℔.	oz.	pwt.	gr.
1	0	15	4
	9	7	5.76
	3	7	22.24, *Ans.*

(7.)

℥.	℈.	gr.
4	1	$1\frac{9}{11}$
	2	$6\frac{2}{13}$
3	1	$15\frac{95}{143}$, *Ans.*

(8.)

T.	cwt.	qr.	lb.
75	0	0	8
56	9	1	23
18	10	2	10, *Ans.*

(9.)

$\frac{5}{24}$ ℔. tr. = 1,200 gr.
$\frac{6}{35}$ lb. av. = 1,200 gr.
0, *Ans*

(10.)

gal.	qt.	pt.	gi.
31	0	1	2
12	1	0	3
18	3	0	3, *Ans.*

(11.)

bu.	pk.	qt.	pt.
	3	5	1
.0625	=	2	
	3	3	1, *Ans.*

(12.)

f℥.	fʒ.	♏︎.
4	2	0
1	4	38
2	5	22, *Ans.*

(13.)

yr.	d.	hr.	min.	sec.
2	175	21	42	20.16
	275	9	12	59
1	$265\frac{1}{4}$	12	29	21.16,
1	265	18	29	21.16, *Ans.*

(14.)

yr.	mo.	da.
1856	7	1
1855	9	22
	9	9, *Ans.*

(15.)

yr.	mo.	da.
1822	4	1
1814	12	31
	7	3, *Ans.*

(16.)

°	′	″
90	0	0
43	18	57.18
46	41	2.82, *Ans.*

(17.)

°	′	″
180	0	0
161	34	11.8
18	25	48.2, *Ans.*

(18.)

ct.	m.
9	$3\frac{3}{4}$
	$9\frac{1}{4}$
8	$4\frac{1}{2}$, *Ans*

(19.)

$	ct.	m.
12	6	$8\frac{1}{3}$
5	43	$2\frac{1}{2}$
6	63	$5\frac{5}{6}$, *Ans.*

(20.)

£	s.	d.
20	0	0
9	18	$6\frac{1}{2}$
10	1	$5\frac{1}{2}$, *Ans.*

(21.)

sq. rd.	sq. yd.	sq. ft.	sq. in.
80	0	0	10
79	30	2	30
			16, *Ans.*

EXPLANATION.—Increasing the first term of the minuend by $\frac{1}{4}$ sq. ft., we take 30 sq. in. from 46 sq. in., leaving 16 sq. in.; then having carried $\frac{1}{4}$ to 2, we increase the second term of the minuend by $\frac{1}{4}$ sq. yd., and subtract $2\frac{1}{4}$ from $2\frac{1}{4}$ leaving 0; then, increasing 30 by $\frac{1}{4}$, we find a *unit* to be the next necessary increase, and, proceeding then as usual, the answer is found, simply, 16 sq. in.

(22.)

sq. rd.	sq. yd.	sq. ft.	sq. in.
3	0	1	1
1	30	1	140
1	0	1	41, *Ans.*

EXPLANATION. — Increase the first term of the minuend by $1\frac{1}{4}$ sq. ft.; and say, "140 sq. in. from 181 sq. in. leave 41 sq. in.;" then carrying $1\frac{1}{4}$ to next term of subtrahend, we have $2\frac{1}{4}$ sq. ft. to be subtracted; increasing the upper by $\frac{1}{4}$ sq. yd., we take $2\frac{1}{4}$ from $3\frac{1}{4}$, leaving 1; then carrying $\frac{1}{4}$ to 30, and using a *unit* increase above, we find the answer, 1 sq. rd., 1 sq. ft., 41 sq. in.

(23.)

rd.	yd.	ft.	in.
3	0	0	2
2	5	1	4
			4, *Ans.*

EXPLANATION.—Let the first increase be $\frac{1}{2}$ ft.; then 4 in. from 8 in. leave 4 in.; the next term of the subtrahend becomes $1\frac{1}{2}$ ft.; and increasing the upper by $\frac{1}{2}$ yd., the remainder there is 0; the next subtrahend term becomes $5\frac{1}{2}$ yd.; a unit of increase answers then to complete the operation.

(24.)

mi.	rd.	ft.	in.
7	0	0	1
4	319	16	3
2	0	0	4, *Ans.*

EXPLANATION.—Let the first increase of the minuend be $\frac{1}{2}$ ft.; the next a unit, and so on, as usual.

(25.)

A.	sq. rd.	sq. yd.	sq. ft.	sq. in.
13	3	0	5	0
11	0	30	8	40
2	1	30	1	32, *Ans.*

EXPLANATION.—The first minuend increase is $\frac{1}{2}$ sq. ft.; at the second minuend term the increase is $\frac{1}{2}$ sq. yd.; at the third minuend term the increase is 2 sq. rd.; and the answer is 2 A. 1 sq. rd. 30 sq. yd. 1 sq. ft. 32 sq. in.

(26.)

A.	sq. rd.	sq. yd.	sq. ft.	sq. in.
18	0	0	3	3
15	3	30	1	142
2	156	$\frac{1}{4}$	1	5, or

2 A. 156 sq. rd. 3 sq. ft. 41 sq. in., *Ans.*

COMPOUND MULTIPLICATION

Art. 223.

(1.)

rd.	ft.	in.
7	10	5
		6
45	$12\frac{1}{2}$	6,

= 45 rd. 13 ft., *Ans.*

(2.)

mi.	rd.	ft.	in.
1	14	8	3
			97
101	126	8	3, *Ans.*

(3.)

sq. yd.	sq. ft.	sq. in.
5	8	106
		13
77	5	82, *Ans.*

(4.)

41 A. 146.1087 sq. rd. = 6706.1087 sq. rd.; and this × 9.046 = 60663.4593 +, sq. rd., or, 379 A. 23.46 sq. rd., nearly. *Ans.*

(5.)

cu. yd.	cu. ft.	cu. in.
10	3	428.15
		67
678	1	1038.05, *Ans.*

(6.)

℔.	oz.	pwt.	gr.
0	7	16	$5\frac{3}{4}$
			174
113	3	5	$16\frac{1}{2}$ *Ans.*

(7.)

℥	ʒ	℈	gr.
0	2	1	13
			20
6	3, *Ans.*		

(8.)

T.	cwt.	qr.	℔.
0	16	1	7.88
			11
8	19	2	11.68, *Ans.*

(9.)

5 gal. 3 qt. 1 pt. 2 gi. = 190 gi.; this × 35.108 = 6670.520 gi.; which = 208 gal. 1 qt. 1 pt. 2.52 gi., *Ans.*

(10.)

bu.	pk.	qt.	pt.
26	2	7	.37
			10
267	0	7	1.7 *Ans.*

(11.)

f℥	fʒ	m.
0	3	48
		12
5	5	36, *Ans.*

(12.)

18 da. 9 hr. 42 m. 29.3 sec. = 1590149.3 sec.; and this × $16\frac{7}{11}$=26454302—, sec.; which = 306 da. 4 hr. 25 m. 2 sec. nearly. *Ans.*

(13.)

£	s.	d.
215	16	$2\frac{1}{4}$
		75
16185	14	$\frac{3}{4}$, *Ans.*

(14.)

10° 28′ 42.5″ = 37722.5″; which × 2.754 = 103887.7650″; hence, 28° 51′ 27.765″, *Ans.*

COMPOUND DIVISION.

Art. 224.

(1.)

	mi.	rd.
7)	16	109
	2	107

(2.)

	rd.	ft.	in.
3)	37	14	11.28
6)	12	10	5.76
	2	1	8.96, *Ans.*

(3.)

C. cu. ft.

83)675 114.66(8 C.

664	
11	
128	
1522.66 cu.ft. (18.3453 + cu. ft.	
83	
692	
664	
286	440
249	415
376	250
332	249
440	1

8 C. 18.3453 + cu. ft., *Ans.*

(4.)

sq. rd.	sq. yd.	sq. ft.	sq. in.
10	29	5	94

$30\frac{1}{4}$

17)$331\frac{1}{2}$ sq.yd. (19 sq. yd.

17	
$161\frac{1}{2}$	2038
153	17
$8\frac{1}{2}$	33
9	17
$81\frac{1}{2}$ sq.ft. (4 sq.ft.	168
68	153
$13\frac{1}{2}$	15
144	
2038 sq. in. ($119\frac{15}{17}$ sq. in.	

19 sq. yd. 4 sq. ft. $119\frac{15}{17}$ sq. in. *Ans.*

(5.)

```
       sq. mi.   A.   sq. rd.
22.5)  6         0    35
       640
       ----
       3840 A.  (170 A.
       225
       ----
       1590
       1575
       ----
         15.0
        160
       ----
       2435 sq. rd. (108 sq. rd.
       225
       -----
        185.0
        1800
       ------
          5.0 ÷ 22½ = 2/9 sq. rd.
       170 A. 108 2/9 sq. rd., Ans.
```

(6.) 1245 cu. yd. 24 cu. ft. + 1627 cu. in. = 58129819 cu. in.; this ÷ 11.303 = 5142866.4 + cu. in.; this reduced, = 110 cu. yd. 6 cu. ft. 338.4 + cu. in., *Ans.*

(7.)

	℥	ʒ	℈	gr.
12)	3	7	0	18
		2	1	16½

Ans.

(8.) 600 T. 7 cwt. 86 lb. = 307401216 dr., which ÷ 29.06 gives 10578156 + dr.; reduced, brings 20 T. 13 cwt. 20 lb. 14 oz. 12 + dr., *Ans.*

(9.) 312 gal. 2 qt. 1 pt. 3.36 gi. = 80058.88 *eighths* gi.; divisor = $\frac{581}{8}$; quotient = 137.79 + gi.; this, reduced, = 4 gal. 1 qt. 1.79 + gi., *Ans.*

(10.) 19302 bu. ÷ 6.21½ = 3105.712 bu., nearly; reduced, brings 2 pk. 6 qt. 1½ pt. +, *Ans.*

(11.)

	yr.	da.	hr.	min.	sec.
5)	76	108	2	38	26.18
9)	15	94	16	7	41.236
	1	253	24	27	31.25—

1 yr. 254 da. 27 min. 31.25 — sec., *Ans.*

(12.)

	°	′	″
9)	152	46	2
	16	58	26 8/9

Ans.

LONGITUDE AND TIME.

Art. 226.

(1.) Cincinnati being 84° 29′ 45″ W. and New York 74° 24″ W., the difference 10° 29′ 21″ ÷ 15 = 41 min. 57.4 sec. difference in time; hence, the required time at the more westerly place is 41 min. 57.4 sec. earlier than 6 A. M., or 18 min. 2.6 sec. after 5 o clock A. M., *Ans.*

(2.) The problem presumes that Springfield is the farther west. Diff. of time 58 min. $1\frac{2}{15}$ sec. × 15 = 14° 30′ 17″; this diff. of long. added to Philadelphia long. 75° 9′ 3″ W., makes required long. 89° 39′ 20″ W., *Ans.*

(3.) 1st. Obviously, *noon* at starting; 2d. to *keep* noon for 24 hrs. the rate of travel to the west must be 15° per hr., or 15 × 69.16 = 1037.4 stat. mi., *Ans.*

(4.) Mobile 88° 2′ 28″ W., Chicago 87° 35′ W.; the latter is farther east 27′ 28″; this ÷ 15 = 1 min. $49\frac{13}{15}$ sec., the advance of Chicago time, *Ans.*

(5.) 9 hr. — 7 hr. 13 min. $32\frac{4}{15}$ sec. = 1 hr. 46 min. $27\frac{11}{15}$ sec., by which Halifax time is faster; this diff. × 15 = 26° 36′ 56″, the distance of Halifax, east, which taken from St. Louis long. 90° 12′ 14″ W., leaves required long., 63° 35′ 18″ W., *Ans.*

(6.) The diff. of time 46 min. 58 sec. × 15 = 11° 44′ 30″, diff. of long.; and this added to Detroit long., 83° 3′ W., = required long. 94° 47′ 30″ W., *Ans.*

(7.) Diff. of longitude 309° 3′ ÷ 15 = 20 hr. 36 min. 12 sec. It is this much *later* in Sidney than in Honolulu, or 41 min. 12 sec. after 12 o'clock A. M. Monday.

NOTE.—The first explorers reached Sidney by traveling east, while Honolulu was reached by sailing west, and the time at the two places was fixed by reckoning in contrary directions from Greenwich. Hence we must take the *longer* arc as our basis.

(8.) St. Petersburg 30° 16′ E., N. Orleans 90° 3′ 28″ W.; diff. of longitude 120° 19′ 28″ $\div$ 15 = 8 hr. 1 min. $17\frac{13}{15}$ sec., *Ans.*

(9.) 12 hr. + 1 hr. — 6 hr. 54 min. 34 sec. = 6 hr. 5 min. 26 sec. diff. of time; this $\times$ 15 = 91° 21′ 30″ diff. of longitude; and Buffalo being so much *west* of Rome which is 12° 28′ *east*, the long. of the former is 91° 21′ 30″ — 12° 28′ = 78° 53′ 30″ W., *Ans.*

(10.) St. Helena 5° 42′ W., San Francisco 122° 27′ 49″ W.; diff. of long. 116° 45′ 49″ $\div$ 15 = 7 hr. 47 min. $3\frac{4}{15}$ sec.; hence required time is so much earlier than 6 P. M., that is, 12 min. $56\frac{11}{15}$ sec. after 10 A. M., *Ans.*

(11.) 4 hr. 43 min. 12 sec. difference in *time* $\times$ 15 = 70° 48′ difference in longitude. Hence, when the time at the ship is 4 hr. 43 min. 12 sec. *earlier* than at Greenwich, the longitude of the former is 70° 48′ W., *Ans.*

ALIQUOT PARTS.

Art. 227.

(1.)

	mi.	rd.	ft.	in.
	2	105	6	2
				30
	69	281	3	6
20 min. = $\frac{1}{3}$ hr.		248	7	$6\frac{2}{3}$
5 min. = $\frac{1}{4}$ of 20 min.		62	1	$10\frac{2}{3}$
4 " = $\frac{1}{5}$ "		49	11	$4\frac{14}{15}$
40 sec. = $\frac{1}{6}$ of 4 min.		8	4	$7\frac{37}{45}$
12 " = $\frac{1}{20}$ "		2	7	$11\frac{71}{75}$
	71 mi.	12	4 ft.	$\frac{8}{225}$ in., *Ans.*

(2.)

1 A. cost	\$11.52
	694
	4608
	10368
	6912
694 A. . . .	7994.88
1 R. $=\frac{1}{4}$ A. .	2.88
20 P. $=\frac{1}{2}$ R. .	1.44
2 P. $=\frac{1}{10}$ of 20	.144
True cost	\$7999.344
Sup. cost	8009.344

Error \$10 too much.

(3.)

1 oz. cost	\$15.46
7 lb. 8 oz. $=$	92 oz.
	3092
	13914
	1422.32
16 pwt. $=\frac{8}{10}$ oz.	12.368
10 gr. $=\frac{1}{48}$ oz.	.322+
1 gr. $=\frac{1}{10}$ of 10	.032+
	\$1435.042+, *Ans.*

(4.)

1 gal.	\$.375 $=\$\frac{3}{8}$
	88
	33.00
2 qt. $=\frac{1}{2}$ gal.	.1875
1 qt. $=\frac{1}{4}$ "	.09375
1 pt. $=\frac{1}{8}$ "	.04687 +
	33.3281 +,

or \$33.33 —, *Ans.*

(5.)

	97920
	8
	783360
4 hr. $=\frac{1}{6}$ da.	16320
1 hr.	4080
20 min. $=\frac{1}{3}$ hr.	1360
5 min. $=\frac{1}{12}$ hr.	340
30 sec. $\frac{1}{10}$ of 5 min.	34
Ans.	805494

(6.)

	\$8190.50
	17
	5733350
	819050
	139238.50
80 rd. $=\frac{1}{4}$	2047.625
40 rd. $=\frac{1}{8}$	1023.8125
20 rd.	511.9062 +
10 rd.	255.9531
	\$143077.796

per mi.	\$84480
40 rd. $=\frac{1}{8}$	10560
20 rd.	5280
10 rd.	2640
	\$18480
	143077.796
	\$161557.80

nearly, *Ans.*

(7.)

1 cwt.	\$3.75
20 T. = 400 "	1500.00
10 lb = $\frac{1}{10}$ "	.375
2 lb. = $\frac{1}{5}$ of 10	.075
1 lb.	.0375
Ans.,	\$1504.2375.

(8.)

	£	s.
	3	6
		17 lb.
	56	2
6 oz. = $\frac{1}{2}$ lb.	1	13
4 " = $\frac{1}{3}$ "	1	2
1 " = $\frac{1}{12}$ "		5.5
16 pwt. = $\frac{1}{5}$ of 4 oz.		4.4
8 gr. = $\frac{1}{60}$ oz.		.091+
1 gr.		.010+
Ans.	£59	7 s. +

(9.)

	16 mi.	$67\frac{3}{8}$ rd. rate per hr.
		598 hr. = 24 da. 22 hr.
	9693	290.25
30 min. = $\frac{1}{2}$ hr.	8	33.6875
20 min. = $\frac{1}{3}$ hr.	5	129.1250
6 min. = $\frac{1}{5}$ of 30	1	198.7375
12 sec. = $\frac{1}{30}$ of 6 min.		17.29125
	9709 mi.	29.09125 rd., *Ans.*

SIMPLE PROPORTION.

Art. 231.

(1.) *Analysis.*—If I can walk $10\frac{1}{2}$ mi. or $\frac{21}{2}$ mi. in 3 hr., I can walk $\frac{1}{3}$ of $\frac{21}{2}$, or $\frac{7}{2}$ mi. in 1 hr.; in 10 hr., 10 times $\frac{7}{2}$ mi., which is 35 mi., *Ans.*

Or, thus; 3 hr. : 10 hr. : : the distance in 3 hr. : the distance in 10 hr.; that is, 3 hr. : 10 hr. : : $10\frac{1}{2}$: $\frac{1}{3}$ of 10 times $10\frac{1}{2}$ mi., or 35 mi., *Ans.*

(2.) 11 ft. 8 in. : 8 ft. 2 in. : : 670 : (?)

140 in. : 98 in. : : 670 : $\frac{670 \times 98}{140} = 469$ times, *Ans.*

(3.) 8 min. 15 sec. : 1 hr. : : 3 min. : (?)

495 sec. : 3600 sec. : : 3 mi. : $\frac{3 \times 3600}{495} = 21\frac{9}{11}$ mi., *Ans.*

(4.) 1 week : 3 wk. 5 da. : : \$1.75 : (?)
7 da. : 26 da. : : 1.75 : $\frac{1.75 \times 26}{7}$ = \$6.50, *Ans.*

(5.) 15 lb. : 132 lb. : : \$5.43$\frac{3}{4}$: \$5.43$\frac{3}{4}$ × 132 ÷ 15 = \$47.85, *Ans.*

(6.) The smaller the rank the larger the file, for the same body; hence,
36 in rank : 42 in rank : : 24 in file : 28, *Ans.*

(7.) 16 sec. : 60 sec. : : 28 times : 28 × 60 ÷ 16 = 105 times, *Ans.*

(8.) Shorter shadow : longer : : smaller height : larger.
2 ft. 2 in. : 25 ft. 9 in. : : 3 ft. 4 in. : (?)
25 in. : 309 in. : : 40 in. : 40 × 309 ÷ 25 = 494.4 in. = 41 ft. 2$\frac{2}{5}$ in., *Ans.*

REMARK.—This simple problem should be illustrated by a diagram, and a number of similar questions should be prepared by the teacher. Attention to this may give the pupil a fair idea of *similarity of triangles* and *proportionality of sides.*

(9.) 160 A. : 840 A. : : \$4.50 : $\frac{450 \times 840}{160}$ = \$2362.50, *Ans.*

(10.) Supposed measure : true measure : : sup. worth : true worth, or, 8 pt. : 7$\frac{1}{2}$ pt. : : \$240 : \$225; and \$240 — \$225 = \$15 gain, *Ans.* Or,
Sup. measure : excess : : sup. worth : gain;
8 : $\frac{1}{2}$: : \$240 : \$15, *Ans.*

(11.) Sup. pound : the lack : : sup. worth : *its* want;
16 oz. : 1$\frac{1}{4}$ oz. : : \$27.52 : \$2.15, *Ans.*

(12.) 1° : 80° 24′ 37″ : : 365000 ft. : (?)
3600″ : 289477″ : : 36500 ft. : 29349751$\frac{7}{18}$ ft., *Ans.*

(13.) 1 da. : 2 hr. 20 min. 5 sec. : : 5000 times : (?)
86400 sec. : 8405 sec. : : 5000 times : 486$\frac{173}{432}$ times, *Ans.*

(14.) *Analysis.*—If it take 108 days of 8$\frac{1}{2}$ or $\frac{34}{4}$ hr. each, it would take 34 times 108 days or 3672 days of $\frac{1}{4}$ hr. each; and for days each 6$\frac{3}{4}$ hr. long, or $\frac{27}{4}$ hr., it would take only $\frac{1}{27}$ so many, or $\frac{1}{27}$ of 3672 days, which is 136 days, *Ans.*

Or thus; The shorter the days the more of them required for the same work; hence, the *required* length being less than the *given*, $6\frac{3}{4}$ hr. : $8\frac{1}{2}$ hr. : : 108 da. : req. no. of da. = 136 da., *Ans.*

(15.) \$1200 : \$1750 : : 20 mon. : $29\frac{1}{6}$ mon. = 2 yr. 5 m. 5 d., *Ans.*

(16.) 20 mon. : 9 mo. : 18 oz. : : $8\frac{1}{10}$ oz., *Ans.*

(17.) The allowance to a man : the saving by a man : : the present number : the additional number, that is, 14 oz. : 6 oz. : : 560 men : 240 men, *Ans.*

(18.) $18\frac{3}{4}$ sec. : 3600 sec. : : 400 ft. : 76800 ft. = $14\frac{6}{11}$ mi., *Ans.*

(19.) 1 ℔ tr. : 1 lb. av. : : 3 £ 6 s. : (?)
144 gr. : 175 gr. : : 66 s. : $80\frac{5}{24}$ s. = 4 £ $2\frac{1}{2}$ d., *Ans.*

(20.) $29\frac{3}{4}$ mi. : 40 mi. : : $12\frac{3}{4}$ da. : $17\frac{1}{7}$ da., *Ans.*

(21.) $\frac{5}{9}$ of a ship being worth \$6000, $\frac{1}{9}$ is worth \$1200, and $\frac{9}{9}$ or the whole is worth \$10800, *Ans.*

(22.) A's tax : B's tax : : A's worth : B's worth,—
\$78.14 : \$256.01 : : \$5840 : \$19133.59, nearly, *Ans.*

(23.) If 16 oz. bring 28 ct., 1 oz. will bring $\frac{1}{16}$ of 28 ct., or $\frac{7}{4}$ ct.; then 4 lb. 6 oz. or 70 oz. will bring 70 times $\frac{7}{4}$ ct., or \$$1.22\frac{1}{2}$, *Ans.*
Thus, also; 16 oz. : 70 oz. : : 28 ct. : \$$1.22\frac{1}{2}$, *Ans.*

(24.) 2 yr. 3 mon. : 5 yr. 6 mo. : : \$160.29 : (?)
27 mo. : 66 mo. : : \$160.29 : \$391.82, *Ans.*

(25.) \$92.54 gain : \$67.32 gain : : \$1156.75 sale : \$841.50 sale, *Ans.*

(26.) \$318.75 worth : \$1285.20 worth : : \$255 cost : \$1028.16 cost, *Ans.*

(27.) Worth of cloth : advance : : worth of flour : *its* advance ; or,

$3.25 per yd. : $3.625 per yd. : : $5.50 : 6.13\frac{6}{13}$, *Ans.*

(28.) 6 mi. ÷ 60 = 528 ft. ; and, 44 ft. : 528 ft. : : 9 strokes : 108 strokes, *Ans.*

Or, thus ; If it be rowed 6 miles in an hour, it is rowed $\frac{1}{60}$ as far, or 528 ft. in 1 minute ; hence, it will take as many times 9 strokes, as 44 ft., the advance by 9 strokes, is contained times in 528 ft., which is 12 times ; hence, 12 times 9, or 108 strokes, *Ans.*

(29.) $100 trade : $847.56 trade : : $7.75 : $65.6859, *Ans.*

(30.) 1 cord : the pile : : cost of cord : (?)

128 : 15 × 10.5 × 12 : : $4.25 : $62.75, *Ans.*

(31.) Fahrenheit has 180° from freezing to boiling ; hence,

180° F. : 1° F. : : 80° R. : $\frac{4}{9}$° R. 180° F. : 1° F. : : 100° C. : $\frac{5}{9}$° C.	*Ans.* 1st.
100° C. : 1° C. : : 80° R. : $\frac{4}{5}$° R. 100° C. : 1° C. : : 180° F. : 1$\frac{4}{5}$° F.	*Ans.* 2d.
80° R. : 1° R. : : 100° C. : 1$\frac{1}{4}$° C. 80° R. : 1° R. : : 180° F. : 2$\frac{1}{4}$° F.	*Ans.* 3d.

(32.) 180° : (108° — 32°) : : 100° C. : 42$\frac{2}{9}$° C.
180° : (108° — 32°) : : 80° R. : 33$\frac{7}{9}$° R. } *Ans.*

(33.) 80° R. : 25° R. : : 180° F. : 56$\frac{1}{4}$° F. ; this + 32° = 88$\frac{1}{4}$°, *Ans.*

80° R. : 25° R. : : 100° C. : 31$\frac{1}{4}$° C., *Ans.*

(34.) 100° C. : 46° C. : : 80° R. : 36$\frac{4}{5}$° R.

100° C. : 46° C. : : 180° F. : 82$\frac{4}{5}$° ; this + 32° = 114$\frac{4}{5}$°, *Ans.*

REMARK.—The teacher who prefers it may use the short method, multiplying by the ratios found in Ex. 31, but the full statements will be good exercise under this article.

(35.) $\frac{2}{3}$ of 180 lb. : 960 lb. : : (?) : 4 ft.

$\frac{1}{2}$ ft. = 6 in., *Ans.*

(36.) 6 ft. 8 in. : 1 ft. 3 in. : : 512 lb. : available power.
80 in. : 15 in. : : 512 lb. : 96 lb.; $\frac{3}{2}$ of 96 lb. = 144 lb., *Ans.*

(37.) 36 in. : $4\frac{1}{2}$ in. : : 1440 lb. : 180 lb.; $\frac{3}{2}$ of 180 lb. = 270 lb., *Ans.*

(38.) 12 ft. : 54 ft. : : $\frac{2}{3}$ of 198 lb. : 594 lb., *Ans.*

(39.) 9 lb. : 4 lb. : : 10 ft. : $4\frac{4}{9}$ ft. = 4 ft. $5\frac{1}{3}$ in., *Ans.*

(40.) Moon's weight : Earth's weight : : Earth's distance : Moon's distance.
123 : 49147 : : 250 miles : 99892 + mi., *Ans.*

(41.) It can be so solved; for, there is a direct ratio of the amounts, 5 A., $13\frac{1}{3}$ A., the time being the same; 5 A. : $13\frac{1}{3}$ A. : : 3 men : (?) *Ans.*

(42.) It can be so solved; for, there is an inverse ratio, — the greater the number of men, the shorter the time for the same labor. It will require one man working 6 times as long as 6 men, that is, 42 days; and 10 men will be kept $\frac{1}{10}$ as long, or, $4\frac{1}{5}$ days.

(43.) It can not be so solved; for, the gains and prices are not in proportion, the cost remaining the same.

(44.) While a true clock indicates that 1440 min. have passed, the losing one indicates but 1435 as having passed in each whole day; hence, the ratio of the indicated lapse to the true lapse, in any period, is 1435 : 1440. From the 6 A. M., on the first, to 11 on the 15th, the time elapsed is $14\frac{5}{24}$ da. Hence,
1435 min. : 1440 min. : : $14\frac{5}{24}$ da. : 14.2578397 da. = 14 da. 6 hr. 11 min. 17.35 + sec. Hence, as the indication is 14 da. 5 hr., the true time is later than 11, by 1 hr. 11 min. 17.35 + sec.; hence, 11 min. 17.35 + sec. past noon, *Ans.*

NOTE.—We see that, *on the 15th P. M.*, it could not again indicate 11 o'clock. As the operator could determine this for himself from the calculation, A. M. was omitted in the statement.

(45.) At the end of 1 hr. the cellar has retained 80—35 or 45 gal.; that is, in cu. ft., $45 \times 231 \div 1728$. Hence,

Am't in 1 hr. : required am't. : : 1 hr. : req. time.

$231 \times 45 \div 1728 : 12 \times 8 \times 6 :: 1$ hr. : (?)

$\frac{12 \times 8 \times 6 \times 1728}{231 \times 45} = 95.75 +$ hr., *Ans.*

(46.) 4 : 9 : : 252 : 567. By the directions we write, 4 : (?) : : 259 : 574. By Prin. 2, the missing term is found $\frac{574 \times 4}{259}$; and dividing by 9, the required multiplier is found $\frac{574 \times 4}{259 \times 9} = \frac{328}{333}$, *Ans.*

(47.) The proportion states that $6 \div 3 = 18 \div 9$. The quotient in each case is 2; and if the addition to each of the four preserves a proportion, the new quotients must also be equal. Now,

1. When any part of a dividend contains a part of the divisor 2 times exactly, the remaining part of the dividend must contain the remaining part of the divisor two times, also, if the *whole dividend* contain the *whole divisor* two times. Here, a part, 6, of the new dividend, contains a part, 3, of the new divisor, 2 times; but the remaining part of the dividend, being the added number, is exactly equal to the remaining part of the divisor, and of course can not contain itself twice;—hence, whatever be the added number the new quotient can never be so great as 2.

2. The dividend and divisor being increased by the *same* number, their difference *after* the increase must be the same that it is now; the new dividend on the left will exceed its divisor by 3, and that on the right will exceed *its* divisor by 9. The new quotients then must be, on the left, 1 *and a fraction whose numerator is* 3; on the right, 1 *and a fraction whose numerator is* 9.

On the left the denominator is 3 *and the increase*, on the right, 9 *and the same increase.*

3. If these fractions be equal, since the right hand numerator is 3 times the left, the right hand denominator must be three times the left; that is, 9 and *once* the increase would have to equal *three times* 3 and three times the increase. But once the increase can not equal three times the increase; hence, the fractions are not equal, the mixed quotients are not equal, the ratios are not the same, and the proportion is not preserved. Q. E. D.

NUMERICAL ILLUSTRATION.—Try the addition of 5. Then $6 : 3 :: 18 : 9$. By addition the ratios become $11 \div 8$ and $23 \div 14$; but $\frac{11}{8} = 1\frac{3}{8}$, and $\frac{23}{14} = 1\frac{9}{14}$. 14 is *not* equal to 3 times 8.

REMARK.—The teacher will have observed the important difference between a mere numerical illustration and a rigid demonstration, and that the former, though useful, can not be substituted for the latter. The strict proof given above is a good exercise for the arithmetician, although, certainly, the algebraist can satisfy himself by a shorter process. Thus:

Let a be the added number. Then, *if* $6 + a : 3 + a :: 18 + a : 9 + a$, $(6 + a)(9 + a)$ must equal $(3 + a)(18 + a)$; or, $54 + 15a + a^2 = 54 + 21a + a^2$, and $15a = 21a$,

Which is impossible. If, instead of "a" we say simply "*the number*," we may translate this demonstration into common language, so that it will be available for the class.

(48.) One fifth is here the *ratio* where the divisor is a unit more than the dividend; *this dividend* is required. If the divisor be 5 times the dividend, for each 1 part *in* that dividend there is an excess of 4 *such* parts in the divisor; then, 4 parts excess : 1 such part :: a unit, the given excess, : dividend required; or, $4 : 1 :: 1 : \frac{1}{4}$, *Ans.*

(49.) The whole will still contain $1\frac{1}{2}$ lb. of salt, and, the required ratio being $\frac{1}{2}$ lb. salt to 40 lb. mixture,

$\frac{1}{2}$ lb. salt : $1\frac{1}{2}$ lb. salt :: 40 lb. mixture : (?)

Then, $\frac{40 \times 3}{2} \div \frac{1}{2} = 120$ lb., the whole; and $120 - 48 = 72$ lb., *Ans.*

COMPOUND PROPORTION.

Art. 233.

(1.)

$$\left.\begin{array}{rcl} 20 \text{ gal.} & : & 6 \text{ gal.} \\ 204 \text{ min.} & : & 136 \text{ min.} \\ 1 \text{ size} & : & 7\frac{1}{2} \text{ sizes} \end{array}\right\} :: 18 \text{ pipes} : (?)$$

$$\frac{18 \times 6 \times 136 \times 7\frac{1}{2}}{20 \times 204 \times 1} = 27 \text{ pipes, } \textit{Ans.}$$

(2.)

$$\left.\begin{array}{ll} 12 \text{ mon.} : & 32 \text{ mon.} \\ \$100 : & \$4500 \end{array}\right\} :: \$8 : (\,?\,)$$

$$\frac{\$8 \times 32 \times 4500}{12 \times 100} = \$960, \textit{ Ans.}$$

(3.)

$$\left.\begin{array}{ll} 14 \text{ men} : & 12 \text{ men} \\ 6 \text{ da.} : & 2 \text{ da.} \\ 25 \text{ A.} : & 80 \text{ A.} \end{array}\right\} :: 10\tfrac{1}{2} \text{ hr.} : (\,?\,)$$

$$\frac{10\frac{1}{2} \text{ hr.} \times 12 \times 2 \times 80}{14 \times 6 \times 25} = 9\tfrac{3}{5} \text{ hr.}, \textit{ Ans.}$$

(4.)

$$\left.\begin{array}{ll} 4 \text{ horses} : & 150 \text{ horses} \\ 15 \text{ cars} : & 1 \text{ car} \end{array}\right\} :: 9 \text{ mi.} : (\,?\,) = \tfrac{150 \times 9}{60} = 22\tfrac{1}{2} \text{ mi.}, \textit{ Ans.}$$

(5.)

$$\left.\begin{array}{ll} 30 \text{ da.} : & 20 \text{ da.} \\ 8 \text{ hr.} : & 10 \text{ hr.} \\ 247.114 \text{ Ha.} : & 197.6912 \text{ Ha.} \end{array}\right\} :: 12 \text{ men} : (\,?\,)$$

$$\frac{20 \times 10 \times 197.6912 \times 12 \text{ men}}{30 \times 8 \times 247.114} = 8 \text{ men}, \textit{ Ans.}$$

(6.)

$$\left.\begin{array}{ll} \$68.75 \text{ profit} : & \$250 \text{ profit} \\ 28 \text{ mon.} : & 8 \text{ mon.} \end{array}\right\} :: \$3750 : (\,?\,)$$

$$\frac{\$3750 \times 250 \times 8}{\$68.75 \times 28} = \$3896.10+, \textit{ Ans.}$$

(7.)

Counting 30 da. = 1 mon., 3 yr. 8 mon. 25 da. = 1345 da.

$$\left.\begin{array}{ll} \$1500 : & \$100 \\ 1345 \text{ da.} : & 360 \text{ da.} \end{array}\right\} :: \$336.25 \text{ worth} : (\,?\,)$$

$$\frac{\$336.25 \times 360 \times 100}{1345 \times 1500} = \$6, \textit{ Ans.}$$

(8.)

$$\left.\begin{array}{lcl} 3500 \text{ men} & : & 1800 \text{ men} \\ 1 \text{ am't} & : & 5 \text{ am'ts} \\ 12 \text{ oz.} & : & 20 \text{ oz.} \end{array}\right\} :: 4\tfrac{1}{2} \text{ mon.} : (\,?\,)$$

$$\frac{4\frac{1}{2} \times 1800 \times 5 \times 20}{3500 \times 12} = \tfrac{135}{7} \text{ mon.} = 1 \text{ yr. } 7\tfrac{2}{7} \text{ mon., } \textit{Ans.}$$

(9.)

$$\left.\begin{array}{lcl} \$4\tfrac{1}{2} \text{ rate} & : & \$7 \text{ rate} \\ 3 \text{ yr,} & : & 1\tfrac{2}{3} \text{ yr.} \end{array}\right\} :: \$540 : (\,?\,)$$

$$\frac{\$540 \times 7 \times 1\frac{2}{3}}{4\frac{1}{2} \times 3} = \$466.66\tfrac{2}{3}, \textit{ Ans.}$$

(10.)

$$\left.\begin{array}{lcl} 18 \text{ long} & : & 7 \text{ long} \\ 1\tfrac{7}{8} \text{ wide} & : & 2\tfrac{1}{2} \text{ wide} \\ 28 \text{ bu.} & : & 120 \text{ bu.} \end{array}\right\} :: 2 \text{ ft. deep} : (\,?\,)$$

$$\frac{2 \times 7 \times 2\frac{1}{2} \times 120}{18 \times 1\frac{7}{8} \times 28} = 4\tfrac{4}{9} \text{ ft., } \textit{Ans.}$$

(11.)

$$\left.\begin{array}{lcl} 100 \text{ sq. in.} & : & 32 \text{ sq. in.} \\ 40 \times 25 \text{ sq. ft.} & : & 75 \times 16 \text{ sq. ft.} \end{array}\right\} :: 4500 \text{ tiles} : (\,?\,)$$

$$\frac{4500 \times 32 \times 75 \times 16}{100 \times 40 \times 25} = 1728 \text{ tiles, } \textit{Ans.}$$

(12.)

$$\left.\begin{array}{lcl} 1\tfrac{1}{2} \text{ thick} & : & 2 \text{ thick} \\ 30 \text{ high} & : & 24 \text{ high} \\ 216 \text{ long} & : & 324 \text{ long} \end{array}\right\} :: 150000 \text{ bricks} : (\,?\,)$$

$$\frac{150000 \times 2 \times 24 \times 324}{1\frac{1}{2} \times 30 \times 216} = 240000 \text{ bricks, } \textit{Ans.}$$

(13.)

$$\left.\begin{array}{lcl} 1 \text{ house} & : & 6 \text{ houses} \\ 16 \text{ long} & : & 18 \text{ long} \\ 12 \text{ wide} & : & 10 \text{ wide} \end{array}\right\} :: 240 \text{ panes} : (\,?\,)$$

$$\frac{240 \times 6 \times 18 \times 10}{16 \times 12} = 1350 \text{ panes, } \textit{Ans.}$$

(14.)

$$\left.\begin{array}{lll} 5000 \text{ copies} & : & 24000 \text{ copies} \\ 8 \text{ folds} & : & 12 \text{ folds} \\ 320 \text{ pp.} & : & 550 \text{ pp.} \end{array}\right\} :: 800 \text{ reams} : (\,?\,)$$

$$\frac{800 \times 24000 \times 12 \times 550}{5000 \times 8 \times 320} = 9900 \text{ reams, } \textit{Ans.}$$

(15.)

$$\left.\begin{array}{lll} 480 \text{ sters} & : & 1152 \text{ sters} \\ 5 \text{ degrees} & : & 2 \text{ degrees} \\ 96 \text{ hr.} & : & 80 \text{ hr.} \\ 3 \text{ (men)} & : & 4 \text{ (boys)} \\ 2 \text{ (at work)} & : & 3 \text{ (hired)} \end{array}\right\} :: 15 \text{ men} : (\,?\,)$$

$$\frac{15 \times 1152 \times 2 \times 80 \times 4 \times 3}{480 \times 5 \times 96 \times 3 \times 2} = 24 \text{ boys, } \textit{Ans.}$$

PERCENTAGE.

Art. 236. CASE I.

FORMULA.—$B \times R = P$.

REMARK.—These solutions, in the main, follow the brief model operations, 1 and 3, p. 189, Higher Arithmetic.

(1.) 1664 men $\times .62\frac{1}{2} =$ 1040 men, *Ans.*

(2.) $\frac{2}{7} \times .35 = \frac{1}{10}$, *Ans.*

(3.) 48 mi. 256 rd. = 15616 rd.; $15616 \times .09\frac{3}{8} = 1464$ rd. = 4 mi. 184 rd., *Ans.*

(4.) $\$3283.47 \times .11\frac{1}{9} = \frac{1}{9}$ of \$3283.47 = \$364.83, *Ans.*

(5.) $33\frac{1}{3}\% = \frac{1}{3}$, and $\frac{1}{3}$ of 127 gal. 3 qt. 1 pt. = 42 gal. 2 qt. 1 pt., *Ans.*

(6.) 98% of 14 cwt. 2 qr. 20 lb. = 1470 lb. $\times .98 =$ 1440.60 lb. = 14 cwt. 1 qr. $15\frac{3}{5}$ lb., *Ans.*

(7.) 40% of 6 hr. 28 min. 15 sec. = 23295 sec. $\times .40$ = 9318.00 sec. = 2 hr. 35 min. 18 sec., *Ans.*

(8.) 75 A. 75 sq. rd. $=$ 12075 sq. rd.; 12075 sq. rd. $\times$ 1.04 $=$ 12558 sq. rd. $=$ 78 A. 78 sq. rd., *Ans.*

(9) 576 pages $\times .15\frac{5}{8} = 90$ pages, *Ans.*

(10.) 144 cattle $\times \frac{9}{16} = 81$ cattle, *Ans.*

(11.) $16\frac{2}{3}\% = \frac{1}{6}$, and $\frac{1}{6}$ of 1932 hogs $=$ 322 hogs, *Ans.*

(12.) $1000\% = 10$; 10 times $\$5.43\frac{3}{4} = \$54.375$, *Ans.*

(13.) $\frac{4}{5} \times \frac{2\frac{1}{2}}{100} = \frac{10}{500} = \frac{1}{50}$, *Ans.*

(14.) $25\% = \frac{25}{100} = \frac{1}{4}$, *Ans.*

(15.) $\frac{18\frac{3}{4}}{100} = \frac{75}{400} = \frac{3}{16}$, *Ans.* $\frac{31\frac{1}{4}}{100} = \frac{125}{400} = \frac{5}{16}$, *Ans.*

$\frac{37\frac{1}{2}}{100} = \frac{75}{200} = \frac{3}{8}$, " $\frac{43\frac{3}{4}}{100} = \frac{175}{400} = \frac{7}{16}$, "

$\frac{56\frac{1}{4}}{100} = \frac{225}{400} = \frac{9}{16}$, " $\frac{62\frac{1}{2}}{100} = \frac{125}{200} = \frac{5}{8}$, "

$\frac{68\frac{3}{4}}{100} = \frac{275}{400} = \frac{11}{16}$, " $\frac{81\frac{1}{4}}{100} = \frac{325}{400} = \frac{13}{16}$, "

$\frac{83\frac{1}{3}}{100} = \frac{250}{300} = \frac{5}{6}$, " $\frac{87\frac{1}{2}}{100} = \frac{175}{200} = \frac{7}{8}$, "

$\frac{93\frac{3}{4}}{100} = \frac{375}{400} = \frac{15}{16}$, "

(16.) $\frac{100}{100} = 1$ time; $\frac{125}{100} = 1\frac{1}{4}$ times; $\frac{250}{100} = 2\frac{1}{2}$ times; $\frac{675}{100} = 6\frac{3}{4}$ times; $\frac{1000}{100} = 10$ times; $\frac{9437.5}{100} = 94\frac{5}{8}$ times the quantity, *Ans.*

(17.) If he sold 40% he sold $\frac{2}{5}$ of his share, and had $\frac{3}{5}$ left; hence, the sale was $\frac{2}{5}$ of $\frac{3}{8} = \frac{3}{20}$, and the remainder was $\frac{3}{5}$ of $\frac{3}{8} = \frac{9}{40}$, *Ans.*

(18.) When he paid 40% or $\frac{2}{5}$ of it, there were $\frac{3}{5}$ remaining; paying 25% or $\frac{1}{4}$ of that balance, he left $\frac{3}{4}$ of the $\frac{3}{5}$, or, $\frac{9}{20}$ of the whole debt; finally, paying 20% or *one fifth* of the second remainder; he now owes *four* fifths of that remainder, or, $\frac{4}{5}$ of $\frac{9}{20}$ of the debt, which is $\frac{9}{25}$ of it, *Ans.*

(19.) 47 gal. 2 qt. 1 pt. = 381 pt.; $6\frac{2}{3}\%$ of 381 pt. = $\frac{1}{15}$ of it = 3 gal. $1\frac{2}{5}$ pt., *Ans.*

(20.) $1200 × .23 = $276 *board;* $1200 × .10$\frac{2}{5}$ = $124.80 *clothing;* $1200 × .06$\frac{3}{4}$ = $81 *books;* $1200 × .00$\frac{7}{12}$ = $7 *newspapers;* $1200 × .12$\frac{7}{8}$ = $154.50 *other expenses;* sum of these = $643.30, and $1200 — $643.30 = $556.70, *saved, Ans.*

(21.) 10% of 20% of $13.50 = $\frac{1}{10}$ of $\frac{1}{5}$ of $13.50 = 27 cts., *Ans.*

(22.) $\frac{2}{5}$ of $\frac{3}{20}$ of $\frac{3}{4}$ = $\frac{9}{200}$; 133\frac{1}{3}$ × $\frac{9}{200}$ = $$\frac{400}{3}$ × $\frac{9}{200}$ = $6, *Ans.*

(23.) If he deducted 3 ct. on each $1, the whole deduction is $\frac{3}{100}$ of $119449, which is $3583.47; as this is made by the allowance of 1 ct. for each cubic foot, the number of cents in this whole deduction is the number of cubic feet, i. e. 358347 cu. ft.; the whole cost of *each* foot is $119449 ÷ 358347 = $.33$\frac{1}{3}$. Since the rough stone cost 16 ct. per foot, the cost of dressing is $33\frac{1}{3}$ ct. — 16 ct. = $17\frac{1}{3}$ ct., *Ans.*

(24.) 40 yr. = $365\frac{1}{4}$ × 40 = 14610 da., and hence he drinks 14610 × 3 = 43830 gi.; .48 times this = 21038.4 gi. = 657 gal. 1 qt. 1 pt. 2.4 gi., *Ans.*

(25.) Having given away 30%, he had 70% left; then giving 20%, or $\frac{1}{5}$ of *this*, he had left $\frac{4}{5}$ of 70%, or 56% of the whole; and, 56% of $1200 = $672. Now, the share of the last and $12 more = that of the third; the last share and $24 more = the second share; the last share and $36 more = the first share; hence, the $672 must be equal to $72 *more* than *four* shares like the last; $600, therefore, = 4 such shares, and *one* = $\frac{1}{4}$ of $600 = $150, *Ans.*

(26.) The increase being $\frac{1}{5}$ of 3.5, is simply .7; but the decrease is $\frac{1}{8}$ of 9.6, or, 1.2; hence, the whole operation diminishes the number by $1.2 - .7$, or, $\frac{1}{2}$; therefore the difference being $3\frac{1}{2}$, and the subtrahend $\frac{1}{2}$. the minuend $=$ $3\frac{1}{2} + \frac{1}{2} = 4$, *Ans.*

Art. 237.

Case II.

$$\text{Formula.}—\frac{P}{B} = R.$$

Remark.—Excepting the examples complicated by fractions, the following are intended to be like the brief operation, p. 191, Higher Arithmetic.

(1.) $.15 \div 2 = 7\frac{1}{2}\%$, *Ans.*

(2.) 2 yd. 2 ft. 3 in. $=$ 99 in.; and 4 rd. $=$ 792 in.; $\frac{99}{792} = \frac{1}{8} = .12\frac{1}{2} = 12\frac{1}{2}\%$, *Ans.*

(3.) 3 gal. 3 qt. $=$ 3.75 gal.; $3.75 \div 31.5 = 11\frac{19}{21}\%$, *Ans.*

(4.) $\frac{2}{3} \div \frac{4}{5} = \frac{2}{3} \times \frac{5}{4} = \frac{5}{6} = .83\frac{1}{3}\%$, *Ans.*

(5.) $\frac{1}{2}$ of $\frac{3}{5}$ of $\frac{4}{7} = \frac{6}{35}$; this $\div \frac{17}{10} = 10\frac{10}{119}\%$, *Ans.*

(6.) $\dfrac{2\frac{1}{2}}{3} \div \dfrac{3\frac{3}{4}}{10} = \dfrac{25}{11\frac{1}{4}} = \dfrac{100}{45} = \dfrac{20}{9} = 222\frac{2}{9}\%$, *Ans.*

(7.) $\$5.12 \div 640 = .008 = .00\frac{4}{5} = \frac{4}{5}\%$, *Ans.*

(8.) $3.20 \div 2000 = .0016 = .00\frac{4}{25} = \frac{4}{25}\%$, *Ans.*

(9.) $750 \div 12000 = \frac{1}{16} = .06\frac{1}{4} = 6\frac{1}{4}\%$, *Ans.*

(10.) 3 qt. $1\frac{1}{2}$ pt. $=$ 7.5 pt.; 5 gal. $2\frac{1}{2}$ qt. $=$ 45 pt.; $7.5 \div 45 = 16\frac{2}{3}\%$, *Ans.*

(11.) A's money has 150 for each 100 of B's; hence, B's is less than A's by $\frac{50}{150}$ or $\frac{1}{3}$ of A's, which is $33\frac{1}{3}\%$ of it, *Ans.*

(12.) $\frac{8}{100}$ of $\frac{35}{100} = \frac{280}{10000} = \frac{2.8}{100} = 2\frac{4}{5}\%$, *Ans.*

(13.) $\frac{2\frac{1}{2}}{100}$ of $\frac{2\frac{1}{2}}{100} = \frac{6\frac{1}{4}}{10000} = \frac{6\frac{1}{4}}{100}$ of $\frac{1}{100} = \frac{1}{16}$ of $\frac{1}{100} = \frac{1}{16}\%$, *Ans.*

(14.) $40\% = \frac{2}{5}$; $62\frac{1}{2}\% = \frac{5}{8}$; $\frac{2}{5}$ of $\frac{5}{8} = \frac{1}{4} = 25\%$, *Ans.*

(15.) 12% of \$75 = \$9; $9 \div 108 = 8\frac{1}{3}\%$, *Ans.*

(16.) 9 parts and 1 part make 10 parts, the whole; and *of* 10, *one* is 10%, *Ans.*

(17.) One yd. = 36 inches; the meter, which is here the base, is (by Art. 209) 39.37043 inches; $36 \div 39.37043 = 91\frac{1729087}{3937043}\%$, *Ans.*

(18.) A 6 mi. square contains 36 sq. mi. = 23040 acres; $9000 \div 23040 = 39\frac{1}{16}\%$, *Ans.*

REMARK.—Where the *problem states* a number of % which is a well known and convenient aliquot, as $20\% = \frac{1}{5}$, $33\frac{1}{3}\% = \frac{1}{3}$, it is not necessary, *in the written operation*, to *repeat* every such statement *as* %; it is often convenient, and always exact, to write *for* the % the equivalent common fraction, the latter being well known; where, however, it is *not* usual or well known, it is better to indicate the steps of the reduction. The following is in point:

(19.) $\frac{2}{5}$ of $\frac{1}{4} = \frac{1}{10} = 10\%$; $\frac{16}{100}$ of $\frac{3}{8} = \frac{6}{100} = 6\%$; $\frac{4\frac{1}{6}}{100}$ of $\frac{120}{100} = \frac{500}{10000} = \frac{5}{100} = 5\%$; $\frac{2}{100}$ of $\frac{8}{10}$ of $\frac{2}{3} = \frac{32}{3000} = \frac{10\frac{2}{3}}{1000} = \frac{1\frac{2}{30}}{100} = 1\frac{1}{15}\%$; $\frac{3}{5}\%$ of 36% of $75\% = \frac{3}{500}$ of $\frac{36}{100}$ of $\frac{3}{4} = \frac{81}{500} \times \frac{1}{100} = \frac{81}{500}\%$; $\frac{6\frac{7}{8}}{100}$ of $\frac{22\frac{1}{2}}{100}$ of $\frac{96}{100} = \frac{55}{500} \times \frac{45}{200} \times \frac{96}{100} = \frac{297}{20000} = \frac{297}{200} \times \frac{1}{100} = 1\frac{97}{200}\%$, *Ans.*

(20.) $30\% = \frac{3}{10}$; $\frac{3}{10} \div \frac{2}{3} = \frac{9}{20} = \frac{45}{100} = 45\%$, *Ans.*

(21.) $\frac{1}{4}$ of $\frac{2}{5} = \frac{1}{10}$; $\frac{1}{10} \div \frac{3}{4} = \frac{4}{30} = 13\frac{1}{3}\%$, *Ans.*

(22.) $7 \div 24 = .29\frac{1}{6} = 29\frac{1}{6}\%$, *Ans.*

Art. 238. CASE III.

FORMULA.—$\frac{P}{R} = B$.

REMARK.—The solutions under this article follow, in the main, the second model, p. 193, Higher Arithmetic. Both methods are used, however, and the work of the class should not be restricted to either.

(1.) \$3.80 ÷ .05 = \$76, *Ans.*

(2.) $\frac{2}{11} \div .80 = \frac{2}{11} \times \frac{100}{80} = \frac{200}{880} = \frac{5}{22}$, *Ans.*

(3.) 16 ÷ .015 = $1066\frac{2}{3}$, *Ans.*

(4.) $31\frac{1}{4}$ ct. ÷ $.15\frac{5}{8}$ = \$$\frac{.31250}{.10625}$ = \$2, *Ans.*

(5) \$10.75 ÷ $.03\frac{1}{3}$ = \$322.50, *Ans.*

(6.) 162 men ÷ .048 = 3375 men. *Ans.*

(7.) \$19.20 = $\frac{6}{10}$%, $\frac{1}{10}$% = \$3.20, 1% = \$32, 100% = \$3200, *Ans.*

(8.) \$189.80 = 104%, 1% = \$1.825, 100% = \$182.50, *Ans.*

(9.) 16 gal. 1 pt. = 129 pt.; 129 pt. ÷ $.06\frac{1}{7}$ = 2100 pt. = $262\frac{1}{2}$ gal. = 262 gal. 2 qt., *Ans.*

(10.) 10 mi. 316 rd. = 3516 rd.; 3516 rd. ÷ .75 = 4688 rd. = 14 mi. 208 rd., *Ans.*

(11.) 36 men = $42\frac{6}{7}$%; 1% = 36 men ÷ $42\frac{6}{7}$ = 36 men × $\frac{7}{300} = \frac{84}{100}$ men; 100% = 84 men, *Ans.*

(12.) 144 sheep = $12\frac{4}{5}$%; 1% = 144 ÷ $12\frac{4}{5}$ = 144 × $\frac{5}{64} = \frac{45}{4}$; 100% = $\frac{45 \times 100}{4}$ = 1125 sheep, *Ans.*

Or, briefly, 144 sheep ÷ .128 = 1125 sheep, *Ans.*

(13.) \$6000 ÷ .35 = \$17142.857+.
\$17142.86, *Ans.*

(14.) 12 pigeons = $2\frac{2}{3}$% of the flock; 1% = 12 × $\frac{3}{8}$ = $\frac{9}{2}$ pigeons; 100% = 450; and 450 — 12 = 438, *Ans.*

(15.) \$10 = $6\frac{1}{4}$%; 1% = \$10 ÷ $6\frac{1}{4}$ = \$1.60; 100% = \$160, *Ans.*

(16.) \$25 $= 62\frac{1}{2}\%$, or, $\frac{5}{8}$ of A's; $\frac{1}{8} =$ \$5, whole $=$ \$40; \$25 $= 41\frac{2}{3}\%$, or, $\frac{125}{300}$ of B's; $\frac{1}{300} = \$\frac{1}{5}$; whole $=$ \$60, *Ans.*

(17.) \$5 $\div .13\frac{1}{3} =$ \$37.50; this $+$ \$5 $=$ \$42.50, *Ans.*

(18.) \$150 $\div$ 48 $=$ \$3.12$\frac{1}{2} = 1\%$; $100\% =$ \$312.50; and, \$312.50 — \$150 $=$ \$162.50, *Ans.*

(19.) 65 A. 106 sq. rd. $=$ 10506 sq. rd.; 10506 sq. rd. $\div$ .03 $=$ 350200 sq. rd. $=$ 2188 A. 120 sq. rd., *Ans.*

(20.) (\$13 $\times$ 12) $\div$ .20 $=$ \$780, *Ans.*

(21.) (40 ct. $\div$ 25) $\div .11\frac{3}{7} = \frac{8}{5}$ ct. $\times \frac{700}{80} =$ 14 ct., *Ans.*

(22.) 81 men $\div$ ($\frac{5}{100}$ of $\frac{60}{100}$) $=$ 81 men $\times \frac{100}{3} =$ 2700 men, *Ans.*

(23.) $7\frac{1}{2}\%$ of $60\% = .045$; \$25000 $\div .045 =$ \$55555.55$\frac{5}{9}$, *Ans.*

(24.) The sister received 2 apples, which were 20% of $37\frac{1}{2}\%$ of $33\frac{1}{3}\%$, or, $\frac{1}{5}$ of $\frac{3}{8}$ of $\frac{1}{3} = \frac{1}{40}$ *of all;* if 2 were *one* 40th, $\frac{40}{40} =$ 80 apples, *Ans.*

(25.) \$3 $\div .31\frac{1}{4} = \frac{1200}{125} =$ \$9.60; \$9.60 $+$ \$3 $=$ \$12.60, *Ans.*

(26.) 8000 bu. $= 57\frac{1}{7}\%$; $1\% =$ 8000 bu. $\times \frac{7}{400} =$ 140 bu.; $100\% =$ 14000 bu.; this — 8000 bu. $=$ 6000 bu., *Ans.*

(27.) 32% of 75% of $800\% = \frac{32}{100}$ of $\frac{3}{4}$ of $8 = 1.92$; $1539 \div 1.92 = 801\frac{9}{16}$, *Ans*

Art. 239. CASE IV.

$$\text{FORMULA.—B} = \begin{cases} A \div (1 + R) \\ D \div (1 - R) \end{cases}$$

REMARK.—In the main, the *short* model operation, first under each problem, has been followed here. The teacher will see the advantage of having the class familiar with both methods.

(1.) \$480 $\div 1.33\frac{1}{3} =$ \$360, *Ans.*

(2.) $\frac{5}{6} \div 1.50 = \frac{5}{9}$, *Ans.*

(3.) 96 da. $\div$ (1 + 100%) = $\frac{1}{2}$ of 96 da. = 48 da., *Ans.*

(4.) 2576 bu. $\div$ (1 — .6) = 2576 bu. $\div$ .4 = 6440 bu., *Ans.*

(5.) $87\frac{1}{2}$ ct. $\div$ (1 — $.87\frac{1}{2}$) = $87\frac{1}{2}$ ct. $\div$ $.12\frac{1}{2}$ = $7, *Ans.*

(6.) 42 mi. 60 rd. = 13500 rd. and 13500 rd. $\div$ (1 —.55) = 30000 rd. = 93 mi. 240 rd., *Ans.*

(7.) 2 lb. $9\frac{29}{96}$ oz. = $41\frac{29}{96}$ oz.; $41\frac{29}{96}$ oz. $\div$ (1 — .5) = twice $41\frac{29}{96}$ oz. = $5\frac{125}{768}$ lb., *Ans.*

(8.) $\frac{7}{12} \div (1 - .99\frac{5}{8}) = \frac{7}{12} \times \frac{800}{3} = 155\frac{5}{9}$, *Ans.*

(9.) $920.93\frac{3}{4}$ $\div$ (1 + $3.37\frac{1}{2}$) = $\frac{\$920.9375}{4.375}$ = $210.50, *Ans.*

(10.) $4358.06\frac{1}{4}$ $\div$ (1 + $2.33\frac{1}{3}$) = $4358.06\frac{1}{4}$ $\div$ $3\frac{1}{3}$ = $1307.41\frac{7}{8}$, *Ans.*

(11.) $64\frac{1}{2}$ gal. = spirit *and* water = $107\frac{1}{2}$% sp.; 64.5 gal. $\div$ 1.075 = 60 gal. spirit, *Ans.*

$7\frac{1}{2}$% of 60 gal. = $4\frac{1}{2}$ gal. water, *Ans.*

(12.) The whole cost of it is expressed in ratio to the cost of the cloth :—

Cost of cloth 100%, Trimmings 30%, *Making* of it 50%. In all $\frac{180}{100}$ of it. If $32 be $\frac{180}{100}$ of the cloth cost, the latter = $32 $\div$ 1.80 = $17.77\frac{7}{9}$; 30% of this = $5.33\frac{1}{3}$; 50 % of it = $8.88\frac{8}{9}$, *Ans.*

(13.) If 1 bu. make $39\frac{1}{5}$ lb. of flour, 80 bu. will make 80 times $39\frac{1}{5}$ lb., or, 3136 lb.; but for each 1 lb. this contains for the farmer, it also contains .04 of a lb. for the miller; hence, it affords the farmer as many lb. as 1.04 is contained times in 3136, which are $\frac{3136}{1.04}$; and $\frac{3136}{1.04}$ lb. = $\frac{3136}{1.04} \div 196 = 15\frac{5}{13}$ bl., *Ans.*

The same result will be had if we deduct the miller's percentage from the 1 bu., or ascertain first the whole number of barrels, and make the deduction from *that*. The pupil should give an analysis corresponding to *each* of these operations:

1st. $39\frac{1}{5}$ lb. $\times$ 80 = 3136 lb.; 3136 $\div$ 196 = 16 bl.; 16 bl. $\div$ 1.04 = $15\frac{5}{13}$ bl., *Ans.*

2d. $(1 \div 1.04) \times 80 \times \frac{196}{5} \div 196 = 15\frac{5}{13}$ bl., *Ans.*

(14.) The number of grains is 455.6538 $\times$ 480, and there will be, in value, as many eagles as 9 pwt. 16.2 gr., or, 232.2 gr. is contained times in the given sum; but for each *one eagle* it contains, it must also contain $.01\frac{1}{2}$ of an eagle as expense; hence, it will bring as many eagles as 1.015 is contained times in the whole number of eagles:

$\frac{455.6538 \times 480}{232.2 \times 1.015} = 928$ eagles, *Ans.*

(15.) $2047 \div (1 - .10 \text{ of } 1.10) = 2047 \div (1 - .11) = 2047 \div .89 = 2300$, *Ans.*

(16.) 6% of 50% of $466\frac{2}{3}$% = $.06 \times \frac{1}{2} \times 4\frac{2}{3} = .14$; hence, by formula, $4246\frac{1}{2} \div 1.14 = 3725$, *Ans.*

(17.) .40 of .50 of .60 of .70 = .084; if .084 were taken out there must have been left .916 of it; this being \$1557.20, the whole must have been \$1557.20 $\div$ .916 = \$1700, *Ans.*

(18.) You had 100% at first; giving away $42\frac{6}{7}$%, you had left $57\frac{1}{7}$%, or, $\frac{4}{7}$ of it, which is \$2; *one*-seventh = $\frac{1}{4}$ of \$2 = \$$\frac{1}{2}$; 7 sevenths, or the whole, = \$$\frac{7}{2}$ = \$3.50, *Ans.*

(19.) 5% *from* the whole leaves 95% *of* the whole; this being 570, 1% is 6, 100% = 600, and 5% of this = 30, *Ans.*

(20.) $33\frac{1}{3}$% *from* the whole leaves $66\frac{2}{3}$%, or $\frac{2}{3}$, of the whole; $\frac{6}{10}$ of the $\frac{2}{3}$ being taken, leaves $\frac{4}{10}$ of the $\frac{2}{3}$, or $\frac{4}{15}$ of the whole; $\frac{3}{4}$ of *this* remainder being taken leaves $\frac{1}{4}$ of it, or $\frac{1}{15}$ of the whole. But this was \$500; if \$500 were $\frac{1}{15}$, the whole was \$7500; $\frac{1}{3}$ of this = \$2500; $\frac{6}{10}$

of remaining \$5000 = \$3000, and \$7500 — (\$3000 + \$500 + \$2500) = \$1500, *Ans.*

(21.) $37\frac{1}{2}\% = \frac{3}{8}$, and $44\frac{4}{9}\% = \frac{4}{9}$. The women are $\frac{4}{9}$ of the men; the children are $\frac{3}{8}$ of $\frac{4}{9}$, or $\frac{1}{6}$ of the men; hence, $\frac{4}{9}$ of the men, $\frac{1}{6}$ of the men, and *once* the men, or, in all, $\frac{29}{18}$ of the number of men make the number in the company, which is 87; *one* 18th = $\frac{1}{29}$ of 87, or 3; the whole = 18 times 3 = 54 *men.* Then, $\frac{4}{9}$ of 54 = 24, the no. of *women,* and $\frac{3}{8}$ of 24 = 9, the no. of *children, Ans.*

(22.) When the stock decreased $33\frac{1}{3}\%$, or $\frac{1}{3}$, there was left $\frac{2}{3}$ of it, and when *this* was decreased 20%, or $\frac{1}{5}$, there remained only $\frac{4}{5}$ of the $\frac{2}{3}$, or $\frac{8}{15}$ of the first value; then, when it rose 20%, or $\frac{1}{5}$, it became $\frac{6}{5}$ of the $\frac{8}{15}$, or $\frac{16}{25}$ of the first value; and, lastly, when the $\frac{16}{25}$ rose $\frac{1}{3}$, it became $\frac{4}{3}$ of $\frac{16}{25}$, or $\frac{64}{75}$ of the first value. Hence, the loss was $\frac{11}{75}$ of the first value; but as this was \$66, *one* 75th was \$6, and the whole was \$450, *Ans.*

(23.) It is plain that:

1st. The factory is worth 96% or $\frac{24}{25}$ of the tannery.

2d. The tannery being worth 16% *more* than the boat, or $\frac{116}{100}$ *of* it, the boat is worth $\frac{100}{116}$, or $\frac{25}{29}$ of the tannery.

3d. 75%, or $\frac{3}{4}$, of the factory must be worth $\frac{3}{4}$ of $\frac{24}{25}$, or $\frac{18}{25}$ of the tannery. Hence, if the boat, or $\frac{25}{29}$ of the value of the tannery, be traded for only $\frac{18}{25}$ of the same value, the loss is equal to $\frac{25}{29} - \frac{18}{25}$, or $\frac{103}{725}$ of the tannery. But this is \$103; hence, *one* 725th is \$1, and the whole is \$725, *Ans.*

PROFIT AND LOSS.

Art. 244. CASE I.

(1.) $\$14.50 \times .14\frac{1}{2} = \$210.25$. *Ans.*

(2.) $\$1760 \times 1.26\frac{1}{4} = \$2222$, *Ans.*

(3.) $42540 × .11$\frac{2}{3}$ = $4963, loss; *Ans.*
$42540 — $4963 = $37577, left, *Ans.*

(4.) $10 × 576 × 1.21$\frac{19}{36}$ = $7000, *Ans.*

(5.) The whole cost is 50 ct. + $\frac{1}{10}$ of 50 ct., in all 55 ct.; the profit being 25% of the investment, is $\frac{1}{4}$ of 55 ct.; hence, selling price = 55 ct. + 13$\frac{3}{4}$ ct. = 68$\frac{3}{4}$ ct., *Ans.*

(6.) Following the formulas for amount and difference,
$5000 × 1.14$\frac{3}{4}$ × 1.08 × (1 — .12) = $5452.92;
this — $5000 = $452.92, *Ans.*

(7.) The sum of the costs = $1.25 + (160 + 160 + 80 + 100 + 100 + 112 + 100) % of $1.25 = 812% + 100% = 912% of $1.25 = $11.40. The receipts = 35 × 70 ct. = $24.50; diff. = $13.10, *Ans.*

(8.) $150 × 1.35 = $202.50, *Ans.*

(9.) Cost = 8 ct. But, *losing* $\frac{1}{10}$, each lb. I have bought enables me to sell *only* .9 of a lb. But, I must realize 1.30 times 8 ct., or 10.4 ct. *on* that .9 of a lb.; hence, selling .9 lb. for 10.4 ct., the price of *one* pound must be 10.4 ct. ÷ .9 = 11$\frac{5}{9}$ ct., *Ans.*

(10.) By the formula for amount; $10000 × 1.20 × 1.20 × 1.20 = $17280, *Ans.*

(11.) Price to be realized, $2.50 × 46 × 1$\frac{1}{4}$ = $143.75; but as this is to be brought by 40 gal., the price of one gal. = $143.75 ÷ 40 = 3.59\frac{3}{8}$, *Ans.*

Art. 245.

CASE II.

REMARK.—The *short* operations here, follow the model, p. 201.

(1.) $4 — $1 = $3, profit; 100% = $1, the base; hence, $3 = 300%, *Ans.*

(2.) $4 — $1 = $3, loss; 100% = $4, the base; and 3 ÷ 4 = 75%, *Ans.*

(3.) 5 parts selling for 9 of the same, is a gain of 4; hence, 4, the gain, ÷ 5, the base, = 80%, *Ans.*

(4.) The first outlay was $125, the second $\frac{6}{10}$ as much, making, in all, $\frac{16}{10}$ of $125, or $200, the whole investment. Of this, $25 came back with the third horse, and $150 by the sale, — in all, $175, leaving a loss of $25. Of the investment, $200, the $25 is 12½%, *Ans.*

(5.) A value decreasing $\frac{1}{4}$ becomes $\frac{3}{4}$ only; if this $\frac{3}{4}$ be increased by $\frac{1}{3}$ of itself, it becomes $\frac{3}{4} + \frac{1}{4}$ = the whole. There was, therefore, no gain, no loss.

(6.) $1728 — $1536 = $192 profit; 192 ÷ 1536 = 12½%, *Ans.*

(7.) The lb. sugar compares with the ℔. Troy as 175 with 144; hence, 175 for 144 is, on the buyer's side, a *loss* of $\frac{31}{175}$, or $17\frac{5}{7}$%; on the grocer's side, a *gain* of $\frac{31}{144}$, or $21\frac{19}{36}$%, *Ans.*

(8.) $2500 — $1750 = $750 loss; 750 ÷ 2500 = $\frac{3}{10}$ = 30%, *Ans.*

(9.) 20% loss leaves 80 of each 100; this 80 increased by 40%, or $\frac{4}{10}$ of itself, becomes 112; hence, receiving 112 for each 100 in the cost, is a gain of 12 on each 100; *i. e.*, 12%, *Ans.*

(10.) For each 100 of the cost, you receive 133⅓ at retail; 10%, or $\frac{1}{10}$, *less than* this is $\frac{9}{10}$ *of* it, or 120; hence, by wholesale, receiving 120 on each 100 is a gain of 20%, *Ans.*

(11.) 15% loss leaves 85 of each 100; but 20% gain on the whole requires the value of 120 for cost of 85; hence, gaining 35 on 85, the rate is 35 ÷ 85 = $41\frac{3}{17}$%, *Ans.*

(12.) 35 ct. + $2.25, amount of tax = $2.60, the total cost; $2.85 — $2.60 = 25 ct. profit; .25 ÷ 2.60 = $9\frac{8}{13}$%, *Ans.*

(13.) Value in hand at first was \$80; outlay exceeding receipt by \$10; the loss 10, is, of the base 80, $12\frac{1}{2}\%$, *Ans.*

Art. 246.

CASE III.

(1.) $\$2000 \div .08 = \$25000$; $\$25000 + 2000 = \$27000$, *Ans.*

(2.) $\$50 \div .225 = \$\frac{2000}{9} = \$222.22\frac{2}{9}$, *Ans.*

(3.) $10 \text{ ct.} \div .13\frac{1}{3} = \frac{300}{4} \text{ ct.} = 75 \text{ ct.}$, *Ans.*

(4.) $\$2\frac{1}{2} \div .07\frac{1}{7} = \$\frac{5}{2} \times \frac{700}{50} = \$35$, *Ans.*

(5.) $\$5 \div .02\frac{7}{9} = \$5 \times \frac{900}{25} = \$180$, A's money;
$\$5 \div .03\frac{1}{3} = \$150 =$ B's; diff. $= \$30$, *Ans.*

(6.) $\$2400 = 120\%$, $1\% = \$20$, $100\% = \$2000$, last year; this being $44\frac{4}{9}\%$, $1\% = \$2000 \div 44\frac{4}{9} = \$45$; $100\% = \$4500$, year before, *Ans.*

(7.) $(40 \div .04\frac{1}{6}) - 40 = 960 - 40 = 920$, the no. sheep, *Ans.*

Art. 247.

CASE IV.

(1.) $\$3.85 \div 1.10 = \$3.50$, *Ans.*

(2.) $\$5 \div 1.33\frac{1}{3} = \$3.75$, *Ans.*

(3.) $\$952.82 \div (1 - .12) = \$1082.75$; $\$1082.75 \times 1.12 = \$1212.68$, *Ans.*

(4.) $\$238 \div (1 - .2) = \$297.50$, the second cost, or first proceeds; and $\$297.50 \div 1.40 = \$212.50$, *Ans.*

(5.) 1st cost $\times 1.13\frac{1}{3} \times 1.24 =$ 2d proceeds; hence, $\frac{4.216}{3}$ of first cost $= \$3952.50$; and $\$3952.50 \times 3 \div 4.216 = \$2812.50$, first cost; $113\frac{1}{3}\%$ of this $= \$3187.50$, *Ans.*

(6.) Cost on delivery must have been 1.08 times the invoice; and proceeds = $1.16\frac{2}{3}$ times 1.08 times, or, 1.26 times invoice; then, \$1260 ÷ 1.26 = \$1000, *Ans.*

(7.) Increasing a value 100% makes twice the value; hence, the value was doubled 6 times, and the final value was $2 \times 2 \times 2 \times 2 \times 2 \times 2$ times, or 64 times, the first, which must, therefore, have been $\frac{1}{64}$ of \$100000, or \$1562.50, *Ans.*

STOCKS AND BONDS.

Art. 250. CASE I.

(1.) $\$50 \times 18 \times .07\frac{1}{2} = \$67.50$, *Ans.*

(2.) $\$50 \times 147 \times .05 = \$367.50$ dividend: \$360.50 ÷ 50 = 7+; hence, 7 shares and \$367.50 — (7 × \$50) = \$17.50 on another, *Ans.*

(3.) $\$150000 \times .04\frac{1}{2} = \$6750$, *Ans.*

(4.) $\$100 \times 50 \times .18 = \$900$, *Ans.*

(5.) $\$4256000 \times .03\frac{1}{2} = \$148960$, *Ans.*

(6.) $(\$75000 \times .07) + \$6500 = \$11750$, *Ans.*

(7.) $(\$25 \times 24 \times .06) \div \$.08 = 450$, the no. of bu., *Ans.*

Art. 251. CASE II.

(1.) $\$324 \div (\$50 \times 72) = .09$ or 9%, *Ans.*

(2.) \$16384.50 ÷ 225000 = .07+; hence, 7%, and \$16384.50 — (225000 × .07) = \$634.50, surplus, *Ans.*

(3.) \$256484 — \$79383 = \$177101 net; this ÷ 3650000 = $4\frac{9}{10}$%, nearly, or $4\frac{1}{2}$%, leaving 177101 — 164250 = \$12851 over, *Ans.*

(4.) $\frac{\$250}{\$100 \times 500} = .00\frac{1}{2} = \frac{1}{2}\%$, *Ans.*

Art. 252.

CASE III.

(1.) $\$18000 \div .15 = \$120000$, *Ans.*

(2.) $(\$94.50 \div .07) \div 50 = 27$ shares, *Ans.*

(3.) $\dfrac{\$50 \times 50 + \$26}{.08} \div 50 = 69$ shares, *Ans.*

Art. 253.

(1.) $\$(50 \times 102 + 15) \div (\$50 \times 1.1) = 93$ shares.

(2.) 1.05 times first stock = stock after first increase. 1.05×1.08 times first stock = second amount. 567 shares $\div (1.05 \times 1.08) = 500$ shares, *Ans.*

PREMIUM AND DISCOUNT.

Art. 256.

CASE I.

NOTE.—Nearly all the solutions under the articles of Premium and Discount are in imitation of the formulas, and thus merely indicate the processes. It will be to the advantage of the class, if the solutions be placed in this manner on the blackboard, and, corresponding to the signs employed, explanations be required in the order of the indicated steps. The problems are easy but the drill they afford is none the less valuable.

(2.) $\$100 \times 18 \times .08 = \$144$; $\$1800 - \$144 = \$1656$, *Ans.*

(3.) $\$1800 \times .04\frac{1}{2} = \$81$; $\$1800 + \$81 = \$1881$: $\$1881 - \$1656 = \$225$, *Ans.*

(4.) $\$50 \times 62 \times 1.28 = \$3968$, *Ans.*

(5.) $\$50 \times 47 \times (1 - .30) = \$1645$, *Ans.*

(6.) $\$150 \times .00\frac{3}{4} = \$1.12\frac{1}{2}$; $\$150 + \$1.12\frac{1}{2} = \$151.12\frac{1}{2}$, *Ans.*

(7.) $\$2568.45 \times 1.005 = \$2581.29+$, *Ans.*

(8.) $\$425 \times .03 = \$12.75$; $\$425 - \$12.75 = \$412.25$, *Ans.*

(9.) $\$5 \times (1 - .06) = \$4.70$, *Ans.*

(10.) $\$50 \times 40 \times (1 - .10) - \$50 \times 32 \times 1.05 = \$120$, *Ans.*

(11.) $\$50 \times 98 \times (1 - .15) - \$4000 \times 1.00\frac{5}{8} = \$140$, *Ans.*

(12.) $\$50 \times 56 \times (.76\frac{1}{2} - .69) = \$210$, *Ans.*

(13.) $\$50 \times 84 \times (1.06 - .91) = \$630$, *Ans.*

(14.) $\$8651.40 \times (1.01\frac{1}{4} - .99\frac{1}{2}) = \$151.399\frac{1}{2}$, or $\$151.40$, *Ans.*

CASE II.

Art. 257.

(1.) $\$2401.30 - \$2360 = \$41.30$ premium; $41.30 \div 2360 = 1\frac{3}{4}\%$, *Ans.*

(2.) $\$50 \times 112 - \$3640 = \$1960$; $1960 \div 5600 = 35\%$, *Ans.*

(3.) $\$5600 \times .08 = \$448$; \$448, the gain, $\div$ \$3640, the investment, $= 12\frac{4}{13}\%$, *Ans.*

(4.) $\$5936 - \$3640 = \$2296$; $2296 \div 3640 = 63\frac{1}{13}\%$; $\$5936 - \$5600 = \$336$; $336 \div 5600 = 6\%$, *Ans.*

(5.) $12\frac{4}{13}\% + 63\frac{1}{13}\% = 75\frac{5}{13}\%$, *Ans.*

(6.) $$\frac{\$50 \times 280 - \$1000 \times 12 \times 1.07}{\$50 \times 280} = 8\frac{4}{7}\%, \textit{ Ans.}$$

(7.) $$\frac{\$266\frac{2}{3} \times (1 - .04) - \$250}{\$250} = 2\frac{2}{5}\%, \textit{ Ans.}$$

(8.) $$\frac{\$50 \times 58 \times 1.40 - \$4000}{\$4000} = 1\frac{1}{2}\%, \textit{ Ans.}$$

(9.) (\$5 — \$4.60) ÷ \$5 = 8%, *Ans.*

(10.) $\frac{\$2600 - (\$2508.03 - \$25.03)}{\$2600} = 4\frac{1}{2}\%$, *Ans.*

CASE III.

Art. 258.

(1.) 36 ct. ÷ $\frac{3}{4}$ = 48 ct. = 1% ; ∴ 100% = \$48, *Ans.*

(2.) \$117 ÷ $2\frac{1}{4}$ × 100 = \$5200 ; \$5200 ÷ \$50 = 104 shares, *Ans.* See *Art.* 86.

(3.) \$93.75 ÷ $7\frac{1}{2}$ × 100 = \$1250 ; \$1250 ÷ \$50 = 25 shares, *Ans.*

(4.) $8\frac{1}{4}\% - 4\frac{1}{2}\% = 3\frac{3}{4}\%$ advance ; \$345 ÷ $3\frac{3}{4}$ × 100 = \$9200 ; \$9200 ÷ \$100 = 92 shares, *Ans.*

(5.) For each \$1 in the par value, the buyer gains $17\frac{1}{2}$ ct. ; then, $\frac{\$192.50 \div .17\frac{1}{2}}{50}$ = 22 shares, *Ans.*

(6.) \$10.36 ÷ $\frac{7}{8}$ × 100 = \$1184, *Ans.*

(7.) 42% — 6% = 36% ; \$666 ÷ 36 × 100 = \$1850 stock ; \$1850 ÷ \$50 = 37 shares, *Ans.*

(8.) Each 90 ct. of my money bought \$1 of the stock, and when the latter became worth \$1.05, the sale brought $\frac{105}{90}$ of the first money. But it required \$1.02 to buy \$1 of the second stock ; hence, the *face* of the stock I *could* have bought, would have been $\frac{100}{102}$ of $\frac{105}{90}$, or $\frac{175}{153}$, of my first money. But I did not obtain so much, having paid out \$33, which would have bought $\$32\frac{6}{17}$ of the second stock. If I could now receive $\$32\frac{6}{17}$, I should have $\frac{175}{153}$ of my first money, but to receive $\$21\frac{6}{17}$ *less*, (or only \$11), would be to have $\frac{153}{153}$ of it ; this can be true only because $\frac{22}{153}$ of the first money = $\$21\frac{6}{17}$; *one* 153d of it is $\frac{1}{22}$ of $\$21\frac{6}{17}$, or $\$\frac{33}{34}$, and $\frac{153}{153}$ of it must be 153 times $\$\frac{33}{34}$, or \$148.50, *Ans.*

CASE IV.

Art. 259.

(1.) $100 + 1\frac{1}{2} = 101\frac{1}{2}$; $\$2861.45 \div 101\frac{1}{2} \times 100 = \$2819.16$, *Ans.*

(2.) $100 - 26 = 74$; $\$1591 \div 74 \times 100 = \$2150$; $\$2150 \div \$50 = 43$ shares, *Ans.*

(3.) $100 - \frac{1}{2} = 99\frac{1}{2}$; $\$6398.30 \div 99\frac{1}{2} \times 100 = \$6430.45$, *Ans.*

(4.) $100 - 65 = 35$; 2% of $35\% = \frac{7}{10}\%$; $35 + \frac{7}{10} = 35.7$; $\$881.79 \div 35.7 \times 100 = \$2470$, *Ans.*

(5.) $\$500 \times 17 = \$8500$; $\$8500 \times 25 \div 100 = \$2125$; $\$8500 - \$2125 = \$6375$; $100 + 6\frac{1}{4} = 106\frac{1}{4}$; $\$6375 \div 106\frac{1}{4} \times 100 = \$6000$; $\$6000 \div \$100 = 60$ shares, *Ans.*

(6.) $100 + \frac{5}{8} = 100\frac{5}{8}$; $\$7567 \div 100\frac{5}{8} \times 100 = \$7520$, *Ans.*

(7.) $\$3172.64 \div (1 - .01\frac{1}{4}) = \$3212.80$, *Ans.*

(8.) $\$100 \times \$54 = \$5400$; $\$5400 \times 12 \div 100 = \$648$; $\$5400 - \$648 = \$4752$; $100 + \frac{1}{4} = 100\frac{1}{4}$; $\$4752 \div 100\frac{1}{4} \times 100 = \$4740.15$, *Ans.*

(9.) $\$1000 \times 72 = \$72000$; $\$72000 \times 6\frac{1}{4} \div 100 = \$4500$; $\$72000 + \$4500 = \$76500$; $100 + 2 = 102$; $\$76500 \div 102 \times 100 = \$75000$; $\$75000 \div \$500 = 150$ bonds, *Ans.*

COMMISSION AND BROKERAGE.

Art. 262. CASE I.

(1.) \$268.40 is the whole collection; 5% of each dollar collected leaves 95 ct., in each dollar, to be paid over, — in all, $\frac{95}{100}$ of \$268.40, which is \$254.98, *Ans.*

(2.) $(\$7.50 \times 650 + \$1.25 \times 35 \times 28) \times .02\frac{1}{4} = \$137.25$, *Ans.*

(3.)

$$\begin{array}{lllll}
36547 \text{ lb.} = \text{whole.} & & & & \\
\underline{16875} \times .06 & \times .03 & = 30.375 & \text{1st.} & \text{com.} \\
19672 & & & & \\
\underline{8246} \times .05 & \times .03 & = 12.369 & \text{2d} & \text{“} \\
11426 \times .05\frac{1}{2} & \times .03 & = \underline{18.8529} & \text{3d} & \text{“} \\
 & & \ 61.59\not{6}+ & & \\
 & \textit{Ans.}, & \$61.60 & \text{whole} & \text{“}
\end{array}$$

(4.) $\$648.75 \times .08 = \$51.90$, the fee; $\$648.75 - \$51.90 = \$596.85$, *Ans.*

(5.) For each \$1 of the debt he accepts 80 ct.; keeping 5% of this, or 4 ct., he pays over 76 ct. for each \$1 of the debt; then, $\$1346.50 \times .76 = \$1023.34$; $\$1346.50 \times .04 = \$53.86$, *Ans.*

(6.) $\$950 + \$575 + \$120 = \$1645$, purchase; $3\frac{1}{3}\%$ of this $= \$54.83$, com.; $\$1645 + \$18.25 + \$36.50 + \$54.83 = \$1754.58$, the whole bill, *Ans.*

(7.) $\$27814.60 \times .03\frac{1}{2} = \$973.51$, *Ans.*

(8.) $\$6231.25 \times (.0275 + .035) = \$389.453$, *Ans.*

(9.) $\$500000 \times .05\frac{1}{2} \times .01\frac{1}{4} = \$343.75$, *Ans.*

(10.) $\$14902.50 \times (.01\frac{1}{4} + .02\frac{7}{8}) = \$614.72\not{8}$

$\$614.73$, *Ans.*

(11.) $\$3850 \times .00\frac{5}{8} = \$26.06\frac{1}{4}$, *Ans.*

(12.) The broker must first take $1\frac{1}{8}\%$ of \$4642.85, or \$52.23+; and I retain $2\frac{1}{2}\% - 1\frac{1}{8}\%$, or $1\frac{3}{8}\%$ of \$4642.85 = \$63.84 nearly, *Ans.*

Case II.

Art. 263.

(1.) $\$50 \div (\$1200 + \$50) = 4\%$, *Ans.*

(2.) $\frac{\$6.43\frac{3}{4} \times 800 - \$5021.25}{\$6.43\frac{3}{4} \times 800} = 2\frac{1}{2}\%$, *Ans.*

This is the same as

$$\frac{\text{amount} - \text{net proceeds}}{\text{amount}} = \frac{\text{commission}}{\text{amount}} = \text{rate.}$$

(3.) \$19017.92 — \$553.92 = \$18464, the cost of building, alone; \$553.92 ÷ \$18464 = 3%, *Ans.*

(4.) Similar to the 3d, and the form is, \$148.72 ÷ (\$5802.57 — \$76.85 — \$148.72) = $2\frac{2}{3}\%$, *Ans.*

(5.) \$52.50 ÷ \$1050 = 5%, *Ans.*

(6.) \$169.20 ÷ \$8460 = 2%, *Ans.*

(7.) \$6.92 ÷ (\$62.28 + \$6.92) = 10%, *Ans.*

(8.) \$38.40 ÷ \$6400 = .006 = $\frac{3}{5}\%$, *Ans.*

(9.) \$24.16 ÷ \$2416 = 1%, brokerage; and (\$24.16 + \$42.28) ÷ \$2416 = $2\frac{3}{4}\%$, com., *Ans.*

Case III.

Art. 264.

(1.) \$3500 ÷ $.02\frac{1}{2}$ = \$140000; \$140000 — \$3500 = \$136500, *Ans.*

(2.) If \$1733.45 = 10% or $\frac{1}{10}$ of the gross receipts, it can only equal $\frac{1}{9}$ of the *net* receipts; \$1733.45 × 9 = \$15601.05, *Ans.*

(3.) Including the expense of packing, the whole commission was \$2376.15 + \$1206.75 = \$3582.90; as this is $1\frac{1}{2}\%$, 1% is \$2388.60, 100% is \$2388.60, the cost; and there were as many lb. as $4\frac{1}{2}$ ct. is contained times in this sum, that is, 5308000 lb., *Ans.* Or, written thus:

(\$2376.15 + \$1206.75) ÷ (\$$.04\frac{1}{2}$ × .015) = 5308000, the number of lb.

(4.) (\$64.05 ÷ $.00\frac{7}{8}$) — \$64.05 = \$7255.95, *Ans.*

(5.) $\$156 \div .01\frac{1}{4} = \$12480$; $\$527.10 + \$156 + \$12480 = \$13163.10$, whole cost; $\$12480 \div 10400 = \$1.20$, cost per bu., *Ans.*

(6.) $2\frac{1}{4}\% - \frac{1}{2}\% = 1\frac{3}{4}\%$; $\$107.03 \div .01\frac{3}{4} = \$6116$, sale; $\$6116 \times .02\frac{1}{4} - \$107.03 = \$30.58$, brokerage; $\$6116 - (\$107.03 + \$30.58) = \$5978.39$, proceeds, *Ans*

CASE IV.

Art. 265.

(1.) $\$207.60 \div (1 - .04) = \$216.25$; $\$216.25 \times .04 = \$8.65$, *Ans.*

(2.) $\$1000 \div 1.025 = \$975.609$, or, \$975.61—, *Ans.*

(3.) $\$539.61 - \$56.85 = \$482.76$, the cost + commission; hence, $\$482.76 \div 1.01\frac{1}{4} = \$476.80$, *Ans.*

NOTE.—Sometimes the *one* rate of the formula, is *made up* of *different* rates. See the two examples following, and those under Art. 270.

(4.) Whole cost $\div$ (1 + rate) = cost of the sugar; $\$1500 \div (1 + .02\frac{1}{4} + .02\frac{1}{2}) = \$\frac{1500}{1.0475} = \$1431.98$, *Ans.*

(5.) Net proceeds $\div$ (1 − rate) = receipts; thus, $\$2448.34 \div (1 - .02\frac{3}{4} - .02\frac{1}{2}) = \$\frac{2448.34}{.94\frac{3}{4}} = \$2584$. Price per lb. $= \$2584 \div 20672 = 12\frac{1}{2}$ ct., *Ans.*

(6.) The agent had 5 ct. out of each \$1 in the receipts, leaving 95 ct. proceeds. But the proceeds consisted of two parts; *one* part, a commission, was $\frac{2}{100}$ of the *other* part, an investment. Hence, if the proceeds were 102 parts, the agent would have 2 of them, that is, his commission was $\frac{2}{102}$ of the proceeds. Therefore, for each \$1 of the receipts the agent had, *first*, 5 ct., *secondly*, $\frac{2}{102}$ of 95 ct.; both commissions making $\$\frac{7}{102}$ for each \$1 of the receipts. Hence, there were as many dollars in the whole as $\$\frac{7}{102}$ is contained times in the whole \$210 commission; $\$210 \div \frac{7}{102} = \$3060$; and whole cost, \$3060 — commission, \$210 = \$2850, cost of sugar, *Ans.*

(7.) Out of each \$1, the agent keeps $3\frac{1}{2}$ ct., puts in flour $66\frac{2}{3}$ ct., and keeps also $1\frac{1}{2}\%$ of $66\frac{2}{3}$ ct.; in all, the outlay is $\$.03\frac{1}{2} + \$.01 + \$.66\frac{2}{3} = \$.71\frac{1}{6}$, leaving out of each \$1, simply $\$.28\frac{5}{6}$; hence, $\$432.50 \div .28\frac{5}{6} = \$1500$, flour; then $3\frac{1}{2}\%$ of this $= \$52.50$, commission; cost of coffee $= \frac{2}{3}$ of $\$1500 = \$1000$; and $1\frac{1}{2}\%$ of $\$1000 = \$15$, commission, *Ans.*

(8.) In each \$1 in the value of the pork, the principal could claim 96 ct., and *of* this, there would be $\frac{\$.96}{1.01\frac{1}{4}}$ to be expended for cotton. [See Formula, and Ex. 6.] Hence, for each $\$\frac{96}{101\frac{1}{4}}$ contained in \$2304, there was \$1 in the value of the pork;

$$\$2304 \div \frac{96}{101\frac{1}{4}} = \$2430, \text{ Pork,}$$
$$\$2430 \times .04 = \$97.20, \text{ 1st Com.,}$$
$$\$2304 \times .01\frac{1}{4} = \$28.80, \text{ 2d Com.}$$

(9.) $\$6.20 \times 1400 = \$8680$; $(\$8680 \times .96 - \$34.16) \div 1.015 = \$8176$, *Ans.*

(10.) On each \$1 in the value of the corn, the agent received 3 ct., and there was a proceeds of 97 ct. This 97 ct. was a sum to be divided into two parts, one of which was 3% of the other, or, simply $\frac{3}{103}$ of the proceeds. Hence, in all, the agent took $\frac{3}{100}$ of the corn, *and* $\frac{3}{103}$ of $\frac{97}{100}$ of it; making together $\frac{291}{10300} + \frac{3}{100}$, or $\frac{6}{103}$ of it; if this were \$12, *one* 103d of it was \$2, and the *whole* \$206, *Ans.*

(11.) The proceeds being $\frac{96}{100}$ of the flour, he lost by commission $\frac{2}{102}$ of that, or $\frac{192}{10200}$ of the flour. Then since the investment *without* the \$4.20, was $\frac{100}{102}$ of the proceeds, he lost $\frac{3\frac{1}{3}}{100}$ of $\frac{100}{102}$ of $\frac{96}{100}$ of the flour, or $\frac{3.2}{102}$ of it. But he lost in the first transaction $\frac{4}{100}$ of the flour; in all, *of the flour*, he lost $\frac{4}{100} + \frac{3.2}{102} + \frac{192}{10200}$, or $\frac{920}{10200}$ of the flour. Of the \$4.20 he lost, first, $\frac{2}{102}$ of it, or $\$\frac{8.40}{102}$; secondly, he lost $\frac{3\frac{1}{3}}{100}$ of $\frac{100}{102}$ of \$4.20 = \$$\frac{14}{102}$; therefore, *on his money*, \$4.20, he lost, \$$\frac{8.40}{102}$ + \$$\frac{14}{102}$ = \$$\frac{22.40}{102}$. Taking this from \$5, we have \$$\frac{487.60}{102}$, the loss, equal to $\frac{920}{10200}$ of the flour; then $\frac{1}{10200}$ of the flour must be \$$\frac{.53}{102}$, and $\frac{10200}{10200}$, or the whole worth of the flour, = \$53, *Ans.*

STOCK INVESTMENTS.

Art. 269. CASE I.

(1.) \$28000 ÷ .70 = \$40000, par value; \$40000 × .08 = \$3200, *Ans.*

(2.) \$100962 ÷ 1.06½ = \$94800, face value; 6% of this = \$5688, income. Premium saved = \$5688 × .00⅛ = \$7.11, *Ans.*

(3.) \$10200 ÷ .30 = \$34000 par; dividend = \$34000 × .06 = \$2040, the income, *Ans.*

(4.) \$36000 ÷ .40 = \$90000 par; 4% of this = \$3600, *Ans.*

(5.) Each \$1 invested in the first buys \$$\frac{100}{106\frac{1}{2}}$, or \$$\frac{200}{213}$, and yields \$$\frac{200}{213}$ × .06½ = \$$\frac{13}{213}$, income; each \$1 in the second buys \$$\frac{100}{104\frac{1}{8}}$, or \$$\frac{800}{833}$, and yields \$$\frac{800}{833}$ × .04½ = \$$\frac{36}{833}$ income. The former exceeds the latter by \$$\frac{3161}{177429}$, which, of a dollar, is $\frac{316100}{177429}$% = $1\frac{138671}{177429}$%, *Ans.*

(6.) The brokerage is estimated on the same base with the premium or discount; hence, in one case the dollar costs 60½ ct. in the other 75½ ct. Hence,

\$(10000 ÷ .605) × .05 = \$826.44625
\$(10000 ÷ .755) × .06 = \$794.70186
5% stock is \$31.744+ better, *Ans.*

Art. 270. CASE II.

(1.) \$46000 — \$56.50 = \$45943.50; and 45943.50 ÷

1.09 = \$42150, par of stock. Income = \$46000 × .05 + \$1072 = \$3372 ; \$3372 ÷ \$42150 = .08, or 8%, *Ans.*

(2.) \$64968.75 × .04$\frac{4}{33}$ = \$2677.50, income ; \$64968.75 ÷ 1.03$\frac{1}{8}$ = \$63000, cost ; \$2677.50 ÷ \$63000 = 4$\frac{1}{4}$%, *Ans.*

(3.) Each \$1 of stock cost, in all, \$1.08$\frac{1}{10}$ + \$.00$\frac{1}{4}$ = \$1.0835 ; hence, the whole stock = \$9850 ÷ 1.0835 = \$9090$\frac{10}{11}$; the income, \$500, ÷ cost, \$9090$\frac{10}{11}$ = 5$\frac{1}{2}$%, *Ans.*

(4.) \$2075 being 5%, 100%, or the whole farm value, = \$41500 ; stock income = \$4100 ; \$1.02 + \$.005 = \$1.025, the cost of \$1 in stock. Proceeds of farm = \$41000 ; \$$\frac{41000}{1.025}$ = par of stock, and \$4100 ÷ \$$\frac{41000}{1.025}$ = 10$\frac{1}{4}$%, *Ans.*

(5.) \$122400 ÷ (1.01$\frac{1}{2}$ + .00$\frac{1}{2}$) = \$120000, par of stock, yielding \$120000 × .04$\frac{1}{6}$ = \$5000 ; \$122500 ÷ (1.03$\frac{1}{2}$ + .00$\frac{1}{2}$) = \$$\frac{122500}{1.04}$; the income being \$2500, the rate = \$2500 ÷ \$$\frac{122500}{1.04}$ = $\frac{1.04}{49}$ = 2$\frac{6}{49}$%, *Ans.*

CASE III.

Art. 271.

(1.) (300 ÷ .04) × .92 = \$6900, *Ans.*

(2) (\$180 ÷ .05) × .75 = \$2700, the money in hand ; (\$180 ÷ .06) × 1.02 = \$3060, cost of state stock ; \$3060 — \$2700 = \$360, *Ans.*

(3.) \$1 of R. R. stock costs 80 ct. ; 6 ct. ÷ 80 ct. = .075 ; \$390 ÷ .075 = \$5200, *Ans.*

(4.) Each \$1 invested in the first buys \$$\frac{10}{9}$, and yields \$$\frac{10}{9}$ × .02$\frac{1}{2}$ = \$$\frac{1}{36}$ income ; each \$1 invested in the second buys \$$\frac{100}{85}$, and yields \$$\frac{100}{85}$ × .03 = \$$\frac{3}{85}$ income. The incomes are as 166$\frac{2}{3}$ and 100 ; hence, the former is $\frac{5}{3}$ of the latter ; that is, it contains \$5 as often as the latter contains \$3. But as the first income is $\frac{1}{36}$ of the investment, for each \$5 in that income there must be \$180 in the investment; and, in like manner, for each \$3 in the second income there are \$85 in the second investment. Then, the incomes being as 5 and 3, the investments must be as 180 and 85, or, the smaller, $\frac{17}{36}$ of the larger; that is, the smaller is $\frac{17}{19}$ *of the difference* between them, and, therefore, equals $\frac{17}{19}$ of \$11400 = \$10200. The larger = \$10200 + \$11400 = \$21600, and both = \$31800, *Ans.*

Also, income of first = $\frac{1}{36}$ of \$21600 = \$600; of the second, $\frac{3}{85}$ of \$10200 = \$360; and the whole income = \$960, *Ans.*

(5.) \$21600 × .99$\frac{3}{8}$ = \$21465, proceeds. The income being \$840, the face of the stock must be \$840 ÷ .06 = \$14000, at 80% costing \$11200, *Ans.*

Also, \$21465 — \$11200 = \$10265, for land; hence, the no. of acres = \$10265 ÷ \$30 = 342$\frac{1}{6}$, *Ans.*

(6.) Each \$1 invested in Phila. 6's, bought \$1 ÷ (1.15$\frac{1}{2}$ + .00$\frac{1}{2}$) = \$$\frac{100}{116}$, and yielded \$$\frac{6}{116}$ = \$$\frac{3}{58}$; each \$1 invested in U. P. 7's bought \$1 ÷ (.89$\frac{1}{2}$ + .00$\frac{1}{2}$) = \$$\frac{10}{9}$, and yielded \$$\frac{7}{90}$. The latter investment was 3 times the former. The former income was $\frac{3}{58}$ of the investment; the latter *would* have been $\frac{7}{90}$ of the *same* investment, but *was* 3 *times* $\frac{7}{90}$, or $\frac{7}{30}$ of it; hence, both incomes were $\frac{3}{58}$ + $\frac{7}{30}$, or $\frac{124}{435}$ of the first investment, which, therefore, was $\frac{435}{124}$ of \$9920 = \$34800, *Ans.* Also, \$34800 × 3 = \$104400, *Ans.*

(7.) Each \$1 paid for 6 per cents bought \$$\frac{100}{115}$, and yielded \$$\frac{6}{115}$; each \$1 paid for 4 per cents bought \$$\frac{100}{112.5}$, and yielded \$$\frac{8}{225}$; in the former case the investment was $\frac{115}{6}$ of the yield, in the latter an *equal* investment would have been $\frac{225}{8}$ of the yield; hence, if the *yields* be *equal*, the costs are as $\frac{115}{6}$ and $\frac{225}{8}$, or as 92 and 135; therefore, the smaller is $\frac{92}{43}$ of their difference, and the larger $\frac{135}{43}$ of that difference; $\frac{92}{43}$ of \$430 = \$920, and $\frac{135}{43}$ of \$430 = \$1350, *Ans.*

Art. 272.

CASE IV.

FORMULA.—$\frac{\text{R. D.}}{\text{M. V.}} = \text{R. I.}$

(1.) By formula, $\frac{6}{30} = 20\%$, *Ans.*

(2.) By formula, $\frac{6}{110} = 5\frac{5}{11}\%$, *Ans.*

(3.) In the first, \$1 buys $\$\frac{100}{99\frac{3}{8}}$, and the income is 4% of this, or $\$\frac{32}{795}$, which is $4\frac{4}{159}\%$ of the investment. In the second, \$1 buys $\$\frac{100}{106}$, and the income is $4\frac{1}{2}\%$ of this, or $\$\frac{9}{212}$, which is $4\frac{13}{53}\%$ of an equal investment. Taking the difference, we find the second better by $\frac{35}{159}\%$, *Ans.*

(4.) Loan brings per yr. $6\frac{1}{2}\%$ of $\$30000 = \$1950$.
Preferred stock " 5% of $\$\frac{30000}{.76} = \$1973.68$
Panama stock " $8\frac{1}{2}\%$ of $\$\frac{30000}{1.25} = \$2040$.
Hence the best is the Panama stock, *Ans.*

(5.) Each \$1 of the stock costs $\$1.05\frac{1}{2} + \$.01\frac{1}{2}$, or \$1.07; the yield of that \$1 is 8 ct., and hence, the income is $\frac{8}{107}$ of the cost, or $7\frac{51}{107}\%$, *Ans.*

Art. 273.

CASE V.

FORMULA.—$\frac{\text{R. D.}}{\text{R. I.}} = \text{M. V.}$

(1.) By formula, $\frac{6}{5} = 120\%$, *Ans.*

(2.) In 1st. \$1 buys $\$\frac{10}{7}$, yielding $\$\frac{10}{7} \times .04 = 5\frac{5}{7}$ ct.
" 2d " " $\$\frac{10}{8}$, " $\$\frac{10}{8} \times .05 = 6\frac{1}{4}$ ct.
" 3d " " $\$\frac{10}{9}$, " $\$\frac{10}{9} \times .06 = 6\frac{2}{3}$ ct.
" 4th " " $\$\frac{10}{12}$, " $\$\frac{10}{12} \times .10 = 8\frac{1}{3}$ ct.
The last yields the highest income.

(3.) By analysis, thus:

4% of the face = 7% of the market value; hence, the face = $\frac{7}{4}$ of the cost; then, \$3430 being $\frac{7}{4}$ of cost, $\frac{1}{4}$ of cost $=$ \$490, and the *whole* cost $=$ 4 times \$490, which is \$1960, *Ans.*

By formula, $\frac{4}{7} \times 3430 = 1960$; ∴ \$1960, *Ans.*

(4.) By formula, $\$\frac{9}{8} = \$1.12\frac{1}{2}$, the market value of \$1; hence, a premium of $12\frac{1}{2}\%$, *Ans.*

(5.) $\$500 \times 2 \div .06\frac{2}{333} = \$16650$, the whole cost.

$\$500 \div .10 = \$5000$, *face* of state stock.

$\$500 \div .04 = \$12500$, *face* of R. R. stock.

The cost of a state stock share $= 120\%$, or $\frac{6}{5}$ the cost of a R. R. share of the same face.

Now, had the R. R. shares been *just so many as* the state stock shares, the *cost* of the R. R. stock would have been $\frac{5}{6}$ of the other cost; but as the R. R. shares were $2\frac{1}{2}$ *times* as many as the others, the cost of R. R. stock was $2\frac{1}{2}$ times $\frac{5}{6}$ of the other cost. Hence, the R. R. investment was $\frac{25}{12}$ of the investment in state stock; the sum of the two investments must have been $\frac{37}{12}$ of the cost of the state stock; hence,

The state stock cost $\frac{12}{37}$ of \$16650 $=$ \$5400, } *Ans.*
" R. R. " " $\frac{25}{37}$ " " $=$ \$11250, }

REMARK.—The teacher should call attention to the *compound ratio* presented in this problem.

Par of state stock to		par of R. R. as		2 :	5.
In value, \$1 "	"	" \$1 "	" "	6 :	5.
Whole value "	"	" whole value "	"	12 :	25.

(6.) By formula, $\frac{5}{9} = 55\frac{5}{9}\%$, *Ans.*

INSURANCE.

Art. 276. CASE I.

(1.) $\frac{5}{8}$ of \$24000 $=$ \$15000; $\$15000 \times .02\frac{1}{4} = \$337.50$; $\frac{2}{3}$ of \$36000 $=$ \$24000; $\$24000 \times .01\frac{1}{8} = \$270$; $\$337.50 + \$270 = \$607.50$, *Ans.*

(2.) (\$2500 + \$600) × .006 = \$18.60, *Ans.*

(3.) \$28000 × .01¾ = \$490, *Ans.*

(4.) \$32760 × .008 = \$262.08, *Ans.*

(5.) \$1800 × .00¾ × 10 = \$135, deposit; \$135 × (1 — .05) = \$128.25, received, *Ans.*

(6.) (\$1275 × .00$\frac{5}{9}$) + \$.75 = \$7.83, *Ans.*

(7.) \$25000 × .009 = \$225; $\frac{4}{5}$% of \$10000 = \$80; \$5000 × .01 = \$50; \$225 — (\$80 + \$50) = \$95, *Ans.*

CASE II.

Art. 277.

(1.) $\frac{2}{3}$ of \$4800 = \$3200; \$19.20 × 100 ÷ \$3200 = $\frac{3}{5}$%, *Ans.*

(2.) \$234 — \$1.50 = \$232.50; \$232.50 × 100 ÷ \$18600 = 1¼%, *Ans.*

(3.) \$46.92 — \$2.50 = \$44.42; \$44.42 × 100 ÷ \$2468 = .018 nearly, or 1$\frac{4}{5}$%, *Ans.*

(4.) \$18000 + \$15000 = \$33000; \$42000 — \$33000 = \$9000; \$18000 × 2½ ÷ 100 = \$450; \$15000 × 3$\frac{4}{5}$ ÷ 100 = \$570; \$9000 × 4$\frac{2}{3}$ ÷ 100 = \$420; \$450 + \$570 + \$420 = \$1440; \$1440 × 100 ÷ \$42000 = 3$\frac{3}{7}$%, *Ans.*

(5.) \$10000 × 3 + \$5000 = \$35000; \$45000 — \$35000 = \$10000; \$262.50 × 100 ÷ \$10000 = 2.625 = 2$\frac{5}{8}$%, *Ans.*

(6.) 2% of $\frac{2}{5}$ is $\frac{4}{5}$%; 2½% of ¼ is $\frac{5}{8}$%; $\frac{4}{5}$% + $\frac{5}{8}$% = $\frac{57}{40}$%; 1½% — $\frac{57}{40}$% = $\frac{3}{40}$%; 1 — $\frac{2}{5}$ — ¼ = $\frac{7}{20}$; for $\frac{7}{20}$ I receive $\frac{3}{40}$% on the whole; for $\frac{1}{20}$, $\frac{1}{7}$ of $\frac{3}{40}$%; for the whole, $\frac{20}{7}$ of $\frac{3}{40}$% = $\frac{3}{14}$%, *Ans.*

CASE III.

Art. 278.

(1.) \$118 ÷ .00$\frac{4}{5}$ = \$14750, *Ans.*

(2.) \$411.375 ÷ .015 = \$27425, *Ans.*

(3.) $(\$42.30 \div .9) \times \frac{8}{5} = \$7520$, *Ans.*

(4.) $\frac{3}{5}$ of $2\frac{1}{2} = \frac{3}{2}$; $2\frac{1}{4}\% - \frac{3}{2}\% = \frac{3}{4}\%$; $\$197.13 \div .00\frac{3}{4} =$ $26284, *Ans.*

(5.) $1\frac{3}{5}\%$ of $\frac{1}{2} = \frac{4}{5}\%$; $1\frac{1}{2}\%$ of $\frac{1}{3}\% = \frac{1}{2}\%$; $\frac{1}{2} + \frac{4}{5} = \frac{13}{10}$; $\frac{8}{5}\% - \frac{13}{10}\% = \frac{3}{10}\%$; $\$58.11 \div .003 = \$19370$, *Ans.*

(6.) $\$10000 \times .02\frac{1}{8} = \$212.50$, and $1\frac{3}{4}\%$ of $\$8000 = \$140$; in all, paid out $352.50; realizing $207.50, my whole receipt must have been $560; $\$560 \div .02 = \$28000$, *Ans.*

(7.) Mutual paid $\frac{2}{3}$ of value + $\frac{3}{2}\%$ of $\frac{2}{12}$ of it $= \frac{803}{1200}$;
" rec'd from Union $\frac{1}{6}$ and from owners
$\frac{7}{4}\%$ of $\frac{2}{3}$ of it, = - - - - - $\frac{214}{1200}$.
" lost the diff., or, $\frac{589}{1200}$ *of value.*

Union *paid* $\frac{1}{6}$ of value, received $\frac{1}{400}$ of value; lost $\frac{197}{1200}$. This is, then, $\frac{392}{1200}$ of value *less* than Mutual lost; hence, $\frac{392}{1200}$ of value $= \$49000$; ∴ value $= \$150000$; owners lost $\frac{1}{3}$ of $\$150000 + \frac{7}{4}\%$ of $\frac{2}{3}$ of it, $= \$51750$, *Ans.*

TAXES.

Art. 281. CASE I.

(1.) $\$486250 \times \frac{78}{100} \div 100 = \$3792.75$, *Ans.*

(2.) $\$3800 \times \frac{96}{100} \div 100 = \$36.48$; $\$36.48 + \$1 = \$37.48$, *Ans.*

(3.)

Tax on	$6000.	is	$75.00
"	800.	"	10.00
"	10.	"	.125
"	5.	"	.062
"	.30	"	.004
		Ans.	$85.19

(4.)

Tax on	$10000.	is	$125.00
"	400.	"	5.00
"	20.	"	.25
"	4.	"	.05
"	.50	"	.006
"	2 polls	"	3.00
		Ans.	$133.31

(5.) The difference = 9 times B's tax.

By table, tax on		\$20000. =	\$250.
"	"	5000. =	62.50
"	"	100. =	1.25
"	"	30. =	.375
"	"	5. =	.0625
			\$314.1875
			9
			\$2827.69—, *Ans.*

Art. 282.

CASE II.

(1.) \$19.53 ÷ \$2604 = $\frac{3}{4}$% = 75 ct. on \$100, *Ans.*

(2.) \$1.25 × 1742 = \$2177.50; \$66913.54 — \$2177.50 = \$64736.04; \$64736.04 ÷ \$6814320 = .95; hence, 95 ct. on \$100, *Ans.*

(3.) \$5670 ÷ \$350000 = $1\frac{31}{50}$%, or \$1.62 on \$100, *Ans.*

(4.) (\$50.46 — \$.150) ÷ \$8704 = $\frac{9}{16}$%, or $56\frac{1}{4}$ ct. on \$100, *Ans.*

Art. 283.

CASE III.

(1.) \$66.96 ÷ $1\frac{4}{5}$ × 100 = \$3720, *Ans.*

(2.) \$564.42 ÷ $\frac{46}{100}$ × 100 = \$122700, *Ans.*

(3.) \$71.61 ÷ $1\frac{8}{25}$ × 100 = \$5425, *Ans.*

(4.) \$4000 — \$1024 = \$2976; \$2976 ÷ $\frac{24}{100}$ × 100 = \$1240000, *Ans.*

(5.) $2\frac{1}{2}$% of 16% = $\frac{2}{5}$%; \$26.04 ÷ $\frac{2}{5}$ × 100 = \$6510, *Ans.*

Art. 284.

CASE IV.

(1.) $100 - 1\frac{7}{20} = 98\frac{13}{20}$; $\$125127.66 \div 98\frac{13}{20} \times 100 =$ \$126840, cap.; \$126840 — \$125127.66 = \$1712.34 tax, *Ans.*

(2.) $100 + 2 = 102$; $\$7599 \div 102 \times 100 = \$7450$, *Ans.*

U. S. REVENUE.

Art. 287.

CASE I.

(1.) $(\$5.65 \times 24 + \$2.25 \times 36) \times .35 = \$75.81$, *Ans.*

(2.) $\$.40 \times 45 \times 36 = \$648$, *Ans.*

(3.) $\$16.50 \times 25 \times .50 = \$206.25$
$\$3 \times 25 = 75.$
\$281.25, *Ans.*

(4.) 6 cwt. 2 qr. 18 lb. = 746 lb.; 36 such boxes weigh 26856 lb.; duty at 2 ct. per lb. \$537.12; ad. val. duty = $\$.02\frac{1}{2} \times .25 \times 26856 = \$167.85$; both = \$704.97, *Ans.*

(5.) $(\$2.56 \times 575 \times .35) + \$.50 \times 1154 = \$1092.20$; $(\$1472 + \$1092.20 + \$160.80) \times 1.15 \div 575 = \$5.45$, *Ans.*

(6.) (112 times $\$.05 \times 20) \times .22\frac{9}{28} = \$25$ per ton, *Ans.*

(7.) Duty on 3724 lb., 10 ct. per lb., = \$372.40
Ad val. 11% on $\$.23 \times 3724$ = 94.217
\$466.617

90% of \$466.617 = \$419.96—, *Ans.*

(8.) $1120 \times 1\frac{1}{4} \times \$.23 = \$322$, value; ad. val. duty = 40% of \$322 = \$128.80; this, with \$112, specific duty, makes \$240.80; whole cost = \$562.80; required sale = 1.25 times \$562.80, or, \$703.50; and $\$703.50 \div 1120 =$ $\$.62\frac{13}{16}$ per yd., *Ans.*

CASE II.

Art. 288.

(1.) $1473.80 ÷ $3684.50 = 40%, *Ans.*

(2.) ($10285.31¼ — $7618.75) ÷ $7618.75 = 35%, *Ans.*

(3.) $.52 × 40 × 63 = $1310.40, value; $.05 × 40 × 63 = $126; $453.60 — $126 = $327.60; ad. val. duty; $327.60 ÷ $1310.40 = ¼ = 25%, *Ans.*

CASE III.

Art. 289.

(1.) $575.80 ÷ .25 = $2303.20; sum = $2879, *Ans.*

(2.) $2970 ÷ 1800 ÷ .60 = $2.75, cost; and, ($2970 + 2.75 × 1800) × 1.20 ÷ 1800 = $5.28, *Ans.*

(3.) $151.20 ÷ .20$\frac{5}{26}$ = $748.80, invoice; 15 × 144 ÷ (151.20 ÷ .35) = 5, bottles to gal.; and $(748.80 + 151.20) × 1.20 ÷ (15 × 144) = 50 ct. per bottle, *Ans.*

CASE IV.

Art. 290.

(1.) $2.50 × 1200 = $3000, spec. duty; $1000 × .60 = $600; these, with $75, make $3675 without invoice and ad. val. duty; $13675 — $3675 = $10000; this being 125% of invoice, the latter is $10000 ÷ 1.25 = $8000 for 100000 cigars, or $80 per thousand, *Ans.*

(2.) $45 — $10 = $35, cost, excluding spec. duty; hence, $35 = 1.25 times invoice, which ∴ is $35 ÷ 1.25 = $28 per ton, *Ans.*

(3.) (6.5 × 3 × 2.8) cu. ft. = 54.6 cu. ft.; at 50 ct. per ft., spec. duty = $27.30, making, with ad. val. given, $48.10, and leaving $81.90 = 1.20 times invoice; hence, invoice = $68.25; $68.25 ÷ 54.6 = $1.25, price per cu. ft., *Ans.*

INTEREST.

Art. 295. CASE I.

FORMULA.—$I = P \times R \times T$.

REMARK.—The model on p. 245, Higher Arithmetic, is followed here; 5 or any greater figure in mill's place is counted 1 ct. in the answers.

1. $\$178.63 \times .07 \times 2\frac{22}{45} = \$31.12$, *Ans.*
2. $\$6084.25 \times .045 \times 1\frac{1}{4} = \$342.24$, *Ans.*
3. $64.30 \times .09 \times 1\frac{157}{180} = \$10.83$, *Ans.*
4. $\$1052.80 \times .10 \times \frac{7}{90} = \$8.19$, *Ans.*
5. $\$419.10 \times .06 \times \frac{32}{45} = \$17.88$, *Ans.*
6. $\$1461.85 \times .10 \times 6\frac{107}{180} = \$964.01$, *Ans.*
7. $\$2601.50 \times .07\frac{1}{2} \times \frac{1}{5} = \$39.02$, *Ans.*
8. $\$8722.43 \times .06 \times 5\frac{1}{2} = \$2878.40$, *Ans.*
9. $\$326.50 \times .08 \times \frac{19}{180} = \$2.76$, *Ans.*
10. $\$1106.70 \times .06 \times 4\frac{31}{360} = \$271.33$, *Ans.*
11. $\$10000 \times .06 \times \frac{1}{360} = \$1.67$, *Ans.*

Art. 297.

(1.)

	1532.45
	.12
	183.8940
	9
	1655.046
2 m. $= \frac{1}{6}$	30.649
6 da. $= \frac{1}{10}$	3.0649
1 da. $= \frac{1}{6}$	.5108
	\$1689.2707, *Ans*

(2.)

	78084.50
	.18
	14055.21
	2
	28110.42
4 m. $= \frac{1}{6}$	4685.07
24 da. $= \frac{1}{5}$	937.014
5 da. $= \frac{1}{24}$	195.211
	\$33927.72, *Ans.*

(3.)

	$512.60
	.07
	35.8820
2 m. = $\frac{1}{6}$	5.9803
6 m. = $\frac{1}{2}$	17.941
18 da. = $\frac{1}{10}$	1.7941
	$25.72, *Ans.*

(4.)

	$1363.20
	.01$\frac{1}{4}$
	34080
	136320
30 da. =	17.0400
6 da. = $\frac{1}{5}$	3.408
3 da. = $\frac{1}{2}$	1.704
	$22.15, *Ans.*

(5.)

	$402.50
	.06
90 d. =	24.15
10 d. = $\frac{1}{9}$	2.683
	$26.83, *Ans.*

(6.)

	$6919.32
	.06
	415.1592
	7
	2906.114
6 m. = $\frac{1}{2}$	207.579
	$3113.69, *Ans.*

(7.)

	$990.73
	.07
	69.3511
6 m. = $\frac{1}{2}$	34.6755
3 m. = $\frac{1}{4}$	17.3377
10 da. = $\frac{1}{9}$	1.9264
9 da. = $\frac{1}{10}$	1.7337
	$55.67, *Ans.*

(8.)

	$4642.68
	.15
	696.4020
4 m. = $\frac{1}{3}$	232.134
1 m. = $\frac{1}{4}$	58.0335
15 da. = $\frac{1}{2}$	29.0167
2 da. = $\frac{1}{60}$	3.8689
	$323.05, *Ans.*

(9.)

	$13024
	.10
	1302.40
6 m. = $\frac{1}{2}$	651.20
3 m. = $\frac{1}{4}$	325.60
10 da. = $\frac{1}{9}$	36.177
3 da. = $\frac{1}{30}$	10.853
	$1023.83, *Ans.*

(10.)

	\$615.38
	.20
	123.076
	4
	492.304
6 m. = $\frac{1}{2}$	61.538
4 m. = $\frac{1}{3}$	41.025
1 m. = $\frac{1}{4}$	10.256
6 d. = $\frac{1}{5}$	2.051
	\$607.17, *Ans.*

(11.)

	\$2066.19
	.30
	619.857
	3
	1859.571
6 m. = $\frac{1}{2}$	309.9285
2 d. = $\frac{1}{90}$	3.4436
	\$2172.94, *Ans.*

(12.)

	\$92.55
	.05
	4.6275
3 m. = $\frac{1}{4}$	1.1568
12 d. = $\frac{1}{30}$	.1542
10 d. = $\frac{1}{9}$	.1285
	\$1.44, *Ans.*

(13.) 117 d. = $\frac{39}{10}$ m.; int. of \$1 = \$.01$\frac{1}{2}$ × $\frac{39}{10}$ = \$.0585; required amount = \$757.35 × 1.9585 = \$801.65, *Ans.*

(14.) Time, 16 m. 21 da.; \$.167 = int. of \$1 at 12%; $\frac{1}{2}$ of \$1883 × .167 = \$157.2305; am't = \$1883 + \$157.23 = \$2040.23, *Ans.*

(15.) Int. of \$1 = \$.01 × $\frac{53}{30}$ = \$.017$\frac{2}{3}$; req. am't = \$262.70 × 1.017$\frac{2}{3}$ = \$267.34, *Ans.*

(16.) Find int. at 12%, then ÷ 12 and × 7$\frac{1}{2}$, thus: \$584.48 × $\frac{.133}{3}$ × $\frac{7.5}{12}$ = \$16.19; am't = \$600.67, *Ans.*

(17.) \$3.9228 = int. for 1 mo. at 1%; \$3.9228 × $\frac{71}{30}$ × $\frac{5}{2}$ = \$23.21, int. at 2$\frac{1}{2}$%; \$392.28 + \$23.21 = \$415.49, *Ans.*

(18.) \$7302.85 × $\frac{.365}{6}$ = \$444.26, *Ans.*

(19.) \$1000000 × $\frac{.07}{2}$ × $\frac{1}{73}$ = \$479.45, *Ans.*

(20.) \$5064.30 × .07 × 7$\frac{2}{5}$ ÷ 12 = \$218.609
\$218.609 ÷ 73 = 2.994
\$215.61, *Ans.*

(21.) Time = 40 m. 4 d.; int. at 4% = $\frac{2}{3}$ of \$12500 × .200$\frac{4}{6}$ = \$1672.22, *Ans.*

(22.) Time = 19 m. 19 d., int. = \$4603.15 $\times \frac{7}{6}$ of $.098\frac{1}{6}$ = \$527.19; am't = 5130.34, *Ans.*

(23.) Time = 93 m. 20 d.; int. = \$13682.45 $\times .468\frac{1}{3} \times \frac{8}{6}$ = \$8543.93, *Ans.*

(24.) Received \$876459.50 $\times$ 2.1 $\times$ $.01\frac{1}{2}$ = \$27608.474
Paid \$106525.20 $\times$ .06 = 6391.512
Gain = \$21216.96
Ans.

(25.) Receives \$100 $\times$ 1.1 $\times$ .02 $\times$ 11 = \$24.20
Pays \$100 $\times$ .06 = 6.
Gain = \$18.20, *Ans.*

(26.) £$493.8\frac{1}{3} \times .06 \times 1\frac{2}{3}$ = £$49.38\frac{1}{3}$ = £49 7s. 8d., *Ans.*

(27.) £$24.93\frac{3}{4} \times .06 \times \frac{10}{12}$ = £$1.2468\frac{3}{4}$ = £1 4s. $11\frac{1}{4}$d., *Ans.*

(28.) £25 $\times .05 \times \frac{21}{12}$ = £$2.18\frac{3}{4}$ = £2 3s. 9d., *Ans.*

(29.) £648.775 $\times .05 \times \frac{176}{365}$ = £15.6417 = £15 12s. 10d., *Ans.*

CASE II.

Art. 298.

(1.) Int. of \$1200 for 1 yr. at 10% = \$120; \$1800 — \$1200 = \$600; \$600 ÷ \$120 = 5. *Ans.* 5 yr.

(2.) \$470.90 — \$415.50 = \$55.40; int. of \$415.50 for 1 yr. at 10% = \$41.55; \$55.40 ÷ \$41.55 = $1\frac{1}{3}$ yr. = 1 yr. 4 mo., *Ans.*

(3.) \$4122.15 — \$3703.92 = \$418.23; int. on \$3703.92 for 1 yr. at 8% = \$296.314; \$418.23 ÷ \$296.314 = 1.41144 yr. = 1 yr. 4 mo. 28 da., *Ans.*

(4.) To double itself it must gain 100%, 100 divided by $4\frac{1}{2}$, 6, $7\frac{1}{2}$, 9, 10, 12, 20, 25, and 30, gives $22\frac{2}{9}$, $16\frac{2}{3}$, $13\frac{1}{3}$, $11\frac{1}{9}$, 10, $8\frac{1}{3}$, 5, 4, and $3\frac{1}{3}$ yr., *Ans.*

(5.) It must gain 200%; 200 divided by 4, 10 and 12, gives 50, 20 and $16\frac{2}{3}$ yr., *Ans.*

(6.) \$1480.78 — \$1374.50 = \$106.28; int. of \$1374.50 for 1 yr. at 10% = \$137.45; \$106.28 ÷ \$137.45 = .7732 yr. = 9 mon. 8 da., *Ans.*

(7.) \$4007.54 — \$3642.08 = \$365.46; int. of \$3642.08 for 1 yr, at 12% = \$437.05; \$365.46 ÷ \$437.05 = .8362 yr. = 10 mon. 1 da., *Ans.*

(8.) Int. of \$175.12 for 1 yr. at 6% = \$10.507; \$6.43 ÷ \$10.507 = .612 yr. = 7 mon. 10 da., *Ans.*

(9.) Int. of \$415.38 for 1 yr. at 7% = \$29.077; \$10.69 ÷ \$29.077 = .3676 yr., × 365 = 134 da., *Ans.*

Art. 299.

Case III.

(1.) Gain = 200 on 100; 200 divided by 5, 10, 15, 20, 25 and 30, will give 40, 20, $13\frac{1}{3}$, 10, 8 and $6\frac{2}{3}$%, *Ans.*

(2.) Gain = 300 on 100; 300 divided by 6, 12, 18, 24 and 30, gives 50, 25, $16\frac{2}{3}$, $12\frac{1}{2}$, 10%, *Ans.*

(3.) Int. of \$35000 for 1 mon. at 1% = \$$29\frac{1}{6}$; \$175 ÷ \$$29\frac{1}{6}$ = 6; hence 6%, *Ans.*

(4.) Int. of \$29200 for 1 da. at 1% = \$$\frac{292}{365}$ = 80 ct.; \$6.40 ÷ 80 ct. = 8; hence 8%, *Ans.*

(5.) Int. of \$12624.80 for 3 mon. at 1% = \$31.562; \$315.62 ÷ \$31.562 = 10; hence 10%, *Ans.*

(6.) 100 — 40 = 60; 5 × 2 = 10, that is 10%, on what cost 60%; $\frac{10}{60} = \frac{1}{6} = .16\frac{2}{3} = 16\frac{2}{3}\%$, *Ans.*

(7.) $\frac{6}{10}\% + \frac{1}{2}\% = 1.1\%$; 1.1% of \$8250 = \$90.75;
\$750 — \$90.75 = \$659.25;
Int. of \$8250 for 1 yr. at 1% = \$82.50;
\$659.25 ÷ \$82.50 = 8—; hence 8%—, *Ans.*

Art. 300.

CASE IV.

(1.) Int. of \$1 for 1 yr. at 6% = 6 ct.; \$1500 ÷ .06 = \$25000, *Ans.*

(2.) Int. of \$1 for 2 yr. 6 mon. at 5% = \$$\frac{1}{8}$; \$1830 ÷ $\frac{1}{8}$ = \$14640, *Ans.*

(3.) Int. of \$1 for 1 mon. at 9% = $\frac{3}{4}$ ct.; \$45 ÷ .00$\frac{3}{4}$ = \$6000, *Ans.*

(4.) Int. of \$1 for 68 da. at 1% a mon. = $\frac{34}{15}$ ct.; \$17 ÷ .00$\frac{34}{15}$ = \$750, *Ans.*

(5.) $\frac{\$656.25}{.03\frac{1}{2} \times \frac{3}{4}}$ = \$25000, *Ans.*

(6.) Int. of \$1 for 9 mon. 11 da. at 10% = $7\frac{29}{36}$ ct.; \$86.15 ÷ .07$\frac{29}{36}$ = \$1103.70, *Ans.*

(7.) Int. of \$1 for 112 da. at 7% = $\frac{98}{45}$ ct.; \$313.24 ÷ .00$\frac{98}{45}$ = \$14383.47, *Ans.*

(8.) Int. of \$1 for 7 mon. 14 da. at 6% = $\frac{56}{15}\%$.; \$146.05 ÷ $\frac{56}{15}\%$. = \$3912.05, *Ans.*

(9.) \$58.78 ÷ (.04 × $1\frac{11}{36}$) = \$1125.57, *Ans.*

(10.) Int. of \$1 for 5 mon. 25 da. at 7% = \$.0340277; \$79.12 ÷ .0340277 = \$2325.16, *Ans.*

CASE V.

Art. 301.

(1.) Am't of \$1 for 2 yr. 3 mon. 12 da. at 6% = \$1.137; \$1367.84 ÷ 1.137 = \$1203.03, *Ans.*

(2.) Am't of \$1 for 10 mon. 26 da. at 10% = \$1.09$\frac{1}{18}$; \$2718.96 ÷ 1.09$\frac{1}{18}$ = \$2493.19, *Ans.*

(3.) Am't of \$1 for 3 yr. 1 mon. 7 da. at 4½% = \$1.13$\frac{77}{80}$; \$4613.36 ÷ 1.13$\frac{77}{80}$ = \$4048.14, *Ans.*

(4.) Am't of \$1 for $\frac{79}{365}$ yr. at 7% = \$1.01$\frac{188}{365}$; \$562.07 ÷ 1.01$\frac{188}{365}$ = \$553.68, *Ans.*

MATURITY OF NOTES.

Art. 303.

(1.) Due, Aug. 5; time, 2 mon. 3 da.; int. of \$560.60 for 2 mon. 3 da., at 7% = \$6.867; \$560.60 + \$6.867 = \$567.47—, *Ans.*

(2.) Due, Aug. 2; time, 6 mon. 3 da.; amount of \$430 at 12% = \$430 × (1 + .061) = \$456.23, *Ans.*

(3.) Due, Nov. 13; time, 3 mon. 3 da.; amount of \$4650.80 at 10% = \$4650.80 × 1.0258⅓ = \$4770.95—, *Ans.*

ANNUAL INTEREST.

Art. 304.

(1.) \$1500 on int. 3 yr. 5 mon. 26 da., draws	\$314.00
\$90 on int. 4 yr. 5 mon. 18 da., draws .	24.12
	1500.
Ans.	\$1838.12

(2.) \$6000 for 6 intervals, 3% an interval,	\$1080.
\$180 for 15 " " "	81.
	6000.
Ans.	\$7161.

(3.) $2500 at 7% from Jan. 11, 1871, to March 17, 1873; 2 yr. 2 mo. 6 da., . . . 382.08\frac{1}{3}$

$175 at 7% for 1 yr. 4 mo. 12 da., . . 16.74$\frac{1}{6}$

2500.

Ans. $2898.825

PARTIAL PAYMENTS.

Art. 307. U. S. Rule.

(1.)

Principal, due Sept. $^{10}/_{13}$, 1882, ...	$304.75
Int. from Sept. 13, 1882, to Nov. 3, 1883, (1 yr. 1 mo. 21 da.), ...	20.88
Amount due Nov. 3, 1883, . . .	$325.63, *Ans.*

(2.)

Principal	$429.30
Int. from Apr. 13, 1873, to Dec. 8, 1873; 7 mo. 25 da.	16.814
Amount	446.114
Payments $10 + $60 = ...	70.
Balance,	$376.11
Int. from Dec. 8, 1873, to July 17, 1874, (7 mo. 9 da.)	13.73
	389.84
Payment, ...	200.
Balance,	$189.84
Int. from July 17, 1874, to Jan. 1, 1875, (5 mo. 15 d.)	5.22
Due Jan. 1, 1875, ...	$195.06, *Ans.*

Remark.—This solution records no surplus interest; but in the following example each deficiency of payment is noted, and the interest found for the days, as on p. 249, Higher Arithmetic.

(3.)

Principal,		$1750.00
Int. to Nov. 25, 1874 (2 yr. 3 da.),		246.02
Amount,		1996.02
Payment,		500.00
	Balance,	$1496.02
Int. to July 18, 1875 (7 mo. 23 da.),		67.78
Payment,		50.00
	Surplus int.,	17.78
Int. to Sept. 1, 1875 (1 mo. 14 da.),		12.80
		1496.02
Amount,		1526.60
Payment,		600.00
	Balance,	$926.60
Int. to Dec. 28, 1875 (3 mo. 27 da.),		21.08
Amount,		947.68
Payment,		75.00
	Balance,	$872.68
Int. to Feb. 10, 1876 (1 mo. 13 da.),		7.30
Due at settlement,		$879.98
		Ans.

(4.)

The periods are 1 yr., 6 mo., 6 mo., and 3 mo. The first payment, exceeding the interest due, was applied according to the rule, and the principal, Apr. 1st, was $306. If the next were applied in like manner it could have been no less than $12.24, the interest of $306 for 6 mo. Hence, *if* the payment equaled the interest, the principal was *no greater* than $306, and so the *amount* at next time of payment no greater than $318.24; and the deduction of $20.40, (exceeding a six months' interest) would have left no greater a principal for the last 3 mo. than $297.84, which would have amounted, at the close, to $303.80, a sum not sufficient. Hence, it can not be true that the last principal was so small, if the payment *was* applied, and consequently it can not be true that the *amount*, previously, was so small as $318.24; but as it could have been *no greater*, *if* payment canceled interest, it follows that that payment *could not* have been applied, and must have been less than $12.24.

Now, that deficient payment must have been applied *with* the $20.40, *at the end of the third interval*, or *at the close*. If the latter, the amount must have been that of $306 for 15 mo., that is, $336.60; and the *sum* of the recorded payments $336.60 — $304.98, or $31.62; but as the interest of $306 through two intervals of 6 mo. each, could not exceed $24.48, such a sum of payments as $31.62, could not have been carried past the third indorsement. Therefore, both of the payments must have been applied when the $20.40 was paid. The amount at the end of 3 mo. being $304.98, the principal must have been $304.98 ÷ 1.02 = $299; the amount of the $306 at the time that balance was left, was $330.48; hence, the sum of payments then applied was $31.48, and we have $31.48 — $20.40 = $11.08, *Ans.*

(5.)

The amount of $175 for 2 yr. being $196, and the difference being $41.60, it is plain that the two payments could not have been held until to-day, for each would have been more than $20, and *that* is more than the interest for one year. Hence, the payment must have exceeded the interest.

For each $1 in the payment applied to the principal, the first time, there would be 6 ct. *less interest* the second time, and consequently 6 ct. *more* paid on the principal the second time; that is, for each $1 paid on the principal the first time, there was paid the second time $1.06, or, both times, $2.06 for each $1 in the first application to the principal; but the *whole* application to the principal is $175 — $154.40 = $20.60; and there were as many dollars in the first application as $2.06 is contained times in $20.60; hence, the first application being $10, and the first interest being $10.50, the payment was $20.50, *Ans.*

Verification.—Principal,		$175.
	One year's int.	10.50
		185.50
	Payment,	20.50
	Balance,	$165.
	One year's int.	9.90
		174.90
		20.50
	Balance,	$154.40

(6.)

It is obvious that the payments must exceed the interest. Each $1 which is applied on the principal at any time diminishes the *interest* for the next year by 10 ct., and hence, 10 ct. more can be applied on the principal, next time; that is, principal is diminished regularly by payments which are 10% more each time. Thus:

For each	$1.	paid on	principal,	1st	time,
There is	1.10	"	"	2d	"
And	1.21	"	"	3d	"
"	1.331	"	"	4th	"
"	1.4641	"	"	5th	"

In all, $6.1051 paid on principal as often as there is $1 in the *first* application to the principal; hence, there are as many dollars in the first such application as $6.1051 is contained times in $2442.04; that is, $400 is the first payment *on the principal;* and as the first interest is $244.204, the required payment is $644.204, *Ans.*

Verification.—	$2442.04		$1762.244
	244.204	3d.	644.204
	2686.244		1118.04
1st.	644.204		111.804
	2042.04		1229.844
	204.204	4th.	644.204
	2246.244		585.64
2d.	644.204		58.564
	1602.04		644.204
	160.204	5th.	644.204
	1762.244		

CONNECTICUT RULE.

Art. 308.

Principal of 2d, ...		$429.30
Int. for 1 yr., ...		25.758
Amount of prin., Apr. 13, 1874, .		455.058
$10 paid Oct. 2, 1873, being less than the interest then due, draws no interest, ...	$10.00	
$60 on interest from Dec. 8, 1873, to Apr. 13, 1874, 4 mo. 5 da., . . .	61.25	71.25
Balance,		$383.808
Int. to Jan. 1, 1875, 8 mo. 19 da., . . .		16.568
		400.376
Amt. of $200 from July 17, 1874, to Jan. 1, 1875, 5 mo. 15 da., . . .		205.50
Balance due,		$194.88 *Ans.*

Principal of 3d, ...	$1750.00
Int. to 1st payment, (2 yr. 3 da.), ...	246.02
	1996.02
Nov. 25, 1874, paid	500.
Balance,	$1496.02
Int. to Nov. 25th, 1875, 1 yr., ...	104.72
	1600.74
$50 payment less than interest then due . .	50.00
Balance,	$1550.74
Am't of $600 from Sept. 1, 1875, to Nov. 25, 1875, 2 mo. 24 da.	609.80
Balance,	$940.94
Int. to Feb. 10, 1876, 2 mo. 16 da., ...	13.904
	954.844
$75 on int. from Dec. 28, 1875, to Feb. 10, 1876, 1 mo. 13 da.	75.627
Balance due,	$879.217 *Ans.*

VERMONT RULE.

Principal, ...	$1480.
Interest to Apr. 12, 1880, (1 yr.), ...	88.80
Am't of $40 from July 25, 1879, to Apr. 12, 1880, (8 mo. 18 da.), ...	41.72
Interest debt, Apr. 12, 1880, . .	$47.08
Int. on $47.08 to Apr. 12, 1881, (1 yr.), . .	2.824
Int. on principal for 1 year, ...	88.80
	$138.704
Am't of $50 from May 20, 1880, to Apr. 12, 1881, (10 mo. 23 da.), ...	52.691
Interest debt, Apr. 12, 1881, . .	$86.013
Int. on $86.013 to Apr. 12, 1882, (1 yr.), . .	5.160
	$91.173
Int. on principal for 1 yr., ...	88.80
	$179.973
$350 on int. from June 3, 1881, to Apr. 12, 1882,	368.025
Balance to apply on principal	188.052
	1480.
	$1291.948
Balance due,	$1291.95, *Ans.*

MERCANTILE RULE.

Art. 309.

(1.)

Principal, $950, due Oct. $^{25}/_{28}$, runs from Jan. 25, in leap year. Hence,

Int. for 277 da. $= \$950 \times .07 \times \frac{277}{366} =$ . . .	$50.329
	950.
Amount,	$1000.329

For'd		$1000.329
Payment, March 2, ...	$225.00	
Int. for 240 da. $= \$225 \times .07 \times \frac{240}{365} =$	10.356	
Payment, May 5, ...	174.190	
Int. for 176 da. $= \$174.19 \times .07 \times \frac{176}{365} =$	5.879	
Payment, June 29, ...	187.500	
Int. for 121 da. $= \$187.50 \times .07 \times \frac{121}{365} =$	4.351	
Payment, for Aug. 1, ...	79.150	
Int. for 88 da. $= \$79.15 \times .07 \times \frac{88}{365} =$	1.335	687.761
	Ans.	$312.57

(2.)

Principal, ...		$600.
Int. from June 12, to Feb. 12, 8 mo., ...		24.00
	Amount,	$624.00
$100 on int. 6 mo. amounts to . . .	$103.	
250 " 3 " "	253.75	
120 " 1 " " . . .	120.60	477.35
	Balance,	$146.65, *Ans.*

TRUE DISCOUNT.

Art. 312.

(1.) Am't of $1, for 3 yr. 5 mo. 20 da., at 7%, = $1.243055 ; $5034.15 ÷ 1.243055 = $4049.82—, and $5034.15 — $4049.82 = $984.33, *Ans.*

(2.) Face = $2500 + int. for 2 yr. 6 mo. 18 da. at 6% (counting days of grace); $2500 × 1.153 = $2882.50 ; and am't of $1 at 8% for 2 yr. 6 mo. 18 da. = $1.204 ; $2882.50 ÷ 1.204 = $2394.10, *Ans.*

BANK DISCOUNT.

Art. 315. Case I.

[The time in days being easily found, the following solutions give only the calculation of proceeds and discount.]

(1.) Int. of \$1 for 138 da, = \$.023; \$792.50 × .023 = \$18.2275, discount; \$792.50 — \$18.227 = \$774.27, proceeds, *Ans.*

(2.) Int. of \$1 for 95 da. = \$1 × .07 × 95 ÷ 360 = \$.018472; \$1962.45 × .018472 = \$36.25, discount; \$1962.45 — \$36.25 = \$1926.20, proceeds, *Ans.*

(3.) Int. of \$1 for 148 da., at 6% = \$.024½; \$2672.18 × .024½ = \$65.47, discount; \$2606.27 — \$65.47 = \$2606.71, proceeds, *Ans.*

(4.) Int. of \$1 for 32 da. = \$.015 × 31 ÷ 30 = \$.0155; \$3886 × .0155 = \$60.23, discount; \$3886 — \$60.23 = \$3823.82, proceeds, *Ans.*

(5.) \$2850 × 1.0405 = \$2965.425, the face to be discounted; int. of \$1 for 182 da. = \$.030⅓; \$2965.425 × .030⅓ = \$89.951, discount; and \$2965.425 — \$89.951 = \$2875.47, proceeds, *Ans.*

(6.) Int. of \$1 for 54 da. = \$.015; \$737.40 × .015 = \$11.06; \$737.40 — \$11.06 = \$726.34, proceeds, *Ans.*

(7.) Int. of \$1 for 144 da. = \$.02; \$4085.20 × .02 = \$81.70, discount; \$4085.20 — \$81.70 = \$4003.50, proceeds, *Ans.*

Art. 316. Case II.

(1.) Bank discount of \$1 for 33 da., at 1½% a mo., = \$.0165; proceeds of \$1 = \$.9835; \$1650 ÷ .9835 = \$1677.68, *Ans.*

(2.) B. disc. of \$1 for 63 da., at 6%, = \$.0105; proceeds of \$1 = \$.9895; \$800 ÷ .9895 = \$808.49, *Ans.*

(3.) B. disc. of \$1 for 93 da., 7%, = $\$.0180\frac{5}{6}$; and \$22.75 ÷ $.0180\frac{5}{6}$ = \$1258.06, *Ans.*

(4.) B. disc. of \$1 for 4 mo. 3 da., at 1% a mo., = \$.041; proceeds = \$.959; \$3375 ÷ .959 = \$3519.29, *Ans.*

(5.) B. disc. of \$1 for 6 mo. 3 da., 10%, = $\$.0508\frac{1}{3}$; proceeds = $\$.9491\frac{2}{3}$; \$4850 ÷ $.9491\frac{2}{3}$ = \$5109.75, *Ans.*

(6.) B. disc. of \$1 for 63 da. = \$.042; proceeds = \$.958; \$768.25 ÷ .958 = \$801.93, *Ans.*

(7.) B. disc. of \$1 for 43 da., 8% = $\$.009\frac{5}{9}$; proceeds = $\$.990\frac{4}{9}$; \$2072.60 ÷ $.990\frac{4}{9}$ = \$2092.60—, *Ans.*

(8.) B. disc. of \$1 for 33 da., 6%, = \$.0055; proceeds = \$.9945; \$1000 ÷ .9945 = \$1005.53, *Ans.*

Also, disc. of \$1 for 93 da. = \$.0155; proceeds = \$.9845; \$1000 ÷ .9845 = \$1015.74, *Ans.*

Case III.

Art. 317.

(1.) Discount of \$100 for 33 da. at 1, $1\frac{1}{4}$, $1\frac{1}{2}$ and 2% a mon. = \$1.10, $\$1.37\frac{1}{2}$, \$1.65 and 2.20; proceeds = \$98.90, $\$98.62\frac{1}{2}$, \$98.35 and \$97.80; int. for 33 da. at 1% a yr. = $\frac{33}{360}$ of 1% = $\frac{11}{12000}$ of principal; \$1.10 ÷ $\frac{11}{12000}$ of \$98.90 = $\frac{12000}{989}$ = $12\frac{132}{989}$%; $\$1.37\frac{1}{2}$ ÷ $\frac{11}{12000}$ of $\$98.62\frac{1}{2}$ = $\frac{12000}{789}$ = $15\frac{55}{263}$%; \$1.65 ÷ $\frac{11}{12000}$ of \$98.35 = $\frac{36000}{1967}$ = $18\frac{594}{1967}$%; \$2.20 ÷ $\frac{11}{12000}$ of \$97.80 = $\frac{4000}{163}$ = $24\frac{88}{163}$%, *Ans.*

(2.) Discount of \$100 for 63 da. at 6, 8 and 10% per annum = \$1.05, \$1.40 and \$1.75; proceeds = \$98.95, \$98.60 and \$98.25; int. for 63 da. at 1% per annum = $\frac{63}{360}$% = $\frac{7}{4000}$ of principal; \$1.05 ÷ $\frac{7}{4000}$ of \$98.95 = $\frac{12000}{1979}$ = $6\frac{126}{1979}$%; \$1.40 ÷ $\frac{7}{4000}$ of \$98.60 = $\frac{4000}{493}$ = $8\frac{56}{493}$%; \$1.75 ÷ $\frac{7}{4000}$ of \$98.25 = $\frac{4000}{393}$ = $10\frac{70}{393}$%, *Ans.*

(3.) Discount of \$100 for 93 da. at 2, $2\frac{1}{2}$ and 3% a mon. = \$6.20, \$7.75 and \$9.30; proceeds = \$93.80, \$92.25 and \$90.70; int. for 93 da. at 1% $= \frac{93}{360}\% = \frac{31}{12000}$ of principal; \$6.20 $\div \frac{31}{12000}$ of \$93.80 $= \frac{12000}{469} = 25\frac{275}{469}\%$; \$7.75 $\div \frac{31}{12000}$ of \$92.25 $= \frac{12000}{369} = 32\frac{64}{123}\%$; \$9.30 $\div \frac{31}{12000}$ of \$90.70 $= \frac{36000}{907} = 39\frac{627}{907}\%$, *Ans.*

(4.) Discount of \$100 for 1 yr. (without grace), at 5, 6, 7, 8, 9, 10 and 12%, is \$5, \$6, \$7, \$8, \$9, \$10 and \$12; proceeds = \$95, \$94, \$93, \$92, \$91, \$90 and \$88; $\frac{5}{95} = .05\frac{5}{19} = .5\frac{5}{19}\%$; $\frac{6}{94} = 6\frac{18}{47}\%$; $\frac{7}{93} = 7\frac{49}{93}\%$; $\frac{8}{92} = 8\frac{16}{23}\%$; $\frac{9}{91} = 9\frac{81}{91}\%$; $\frac{10}{90} = 11\frac{1}{9}\%$; $\frac{12}{88} = 13\frac{7}{11}\%$, *Ans.*

(5.) The note being *legally* due in that time, 1 yr. 4 mon. 20 da. includes the days of grace. Out of each \$1 due in that time he takes a discount of $11\frac{1}{9}$ ct., and, therefore, *pays out* for that \$1, $88\frac{8}{9}$ ct. Hence, he receives $11\frac{1}{9}$ ct. interest on $88\frac{8}{9}$ ct. in $1\frac{7}{18}$ yr., or \$1 on \$8 in $1\frac{7}{18}$ yr.; \$8 at *one* % yields in that time $\$.11\frac{1}{9}$; hence, to yield \$1, the rate % must be $1 \div .11\frac{1}{9}$; hence, 9%, *Ans.*

Art. 318.

CASE IV.

(1.) Int. of \$100 for 33 da. at 10, 15 and 20%, is $\$.91\frac{2}{3}$, $\$1.37\frac{1}{2}$, $\$1.83\frac{1}{3}$; amounts are $\$100.91\frac{2}{3}$, $\$101.37\frac{1}{2}$, $\$101.83\frac{1}{3}$; but int. of any sum for 33 da., at 1%, $= \frac{11}{12000}$ of that sum; and

$$\$.91\tfrac{2}{3} \div \tfrac{11}{12000} \text{ of } \$100.91\tfrac{2}{3} = 9\tfrac{1101}{1211}\%.$$
$$\$1.37\tfrac{1}{2} \div \tfrac{11}{12000} \text{ of } \$101.37\tfrac{1}{2} = 14\tfrac{646}{811}\%.$$
$$\$1.83\tfrac{1}{3} \div \tfrac{11}{12000} \text{ of } \$101.83\tfrac{1}{3} = 19\tfrac{391}{611}\%.$$

(2.) Int. of \$100 for 63 da. at 6, 8, 10% = \$1.05, \$1.40, \$1.75; am't = \$101.05, \$101.40, \$101.75; int. for 63 da. at 1% $= \frac{63}{360}\% = \frac{7}{4000}$ of principal;

$\$1.05 \div \frac{7}{4000}$ of $\$101.05 = \frac{12000}{2021} = 5\frac{1895}{2021}\%$, *Ans.*
$\$1.40 \div \frac{7}{4000}$ of $\$101.40 = \frac{4000}{507} = 7\frac{451}{507}\%$, *Ans.*
$\$1.75 \div \frac{7}{4000}$ of $\$101.75 = \frac{4000}{407} = 9\frac{337}{407}\%$, *Ans.*

(3.) Int. of \$100 for 93 da. at 1, 2, 4% a month = \$3.10, \$6.20 and \$12.40; amount = \$103.10, \$106.20 and \$112.40; int. for 93 da. at $1\% = \frac{93}{360}\% = \frac{31}{12000}$ of principal;

$\$\ 3.10 \div \frac{31}{12000}$ of $\$103.10 = \frac{12000}{1031} = 11\frac{659}{1031}\%$, *Ans.*
$\$\ 6.20 \div \frac{31}{12000}$ of $\$106.20 = \frac{4000}{177} = 22\frac{106}{177}\%$, *Ans.*
$\$12.40 = \frac{31}{12000}$ of $\$112.40 = \frac{12000}{281} = 42\frac{198}{281}\%$, *Ans.*

(4.) Int. of \$100 for 1 yr. = \$5, \$6, \$7, \$8, \$9, \$10; amount = \$105, \$106, \$107, \$108, 109, \$110; then,

$\frac{5}{105} = 4\frac{16}{21}\%$; $\frac{6}{106} = 5\frac{35}{53}\%$; $\frac{7}{107} = 6\frac{58}{107}\%$;
$\frac{8}{108} = 7\frac{11}{27}\%$; $\frac{9}{109} = 8\frac{28}{109}\%$; $\frac{10}{110} = 9\frac{1}{11}\%$.

DOMESTIC EXCHANGE.

Art. 320.

N. B.—Observe the addition of 1 cent for 5 mills or more in Ans.

(1.) $\$3805.40 \times 1.00\frac{1}{2} = \$3824.43$—, *Ans.*

(2.) $\$1505.40 \times (1 - .00\frac{1}{4}) = \$1501.64$, *Ans.*

(3.) $\$2000 \div 1.00\frac{5}{8} = \$1987.58$, *Ans.*

(4.) $\$4681.25 \div .98\frac{3}{4} = \$4740.51$, *Ans.*

(6.) Int. of \$12692.50 for 63 da., 6%, = \$133.27; $\frac{3}{4}\%$ of \$12692.50 = \$95.19; \$12692.50 — \$133.27 + \$95.19 = \$12654.42, *Ans.*

(7.) Int. of \$1 for 21 da. = $\frac{7}{20}$ ct.; \$1 face costs \$1.005 — $\frac{7}{20}$ ct. = \$1.0015; $\$5264.15 \div 1.0015 = \$5256.27$, *Ans.*

(8.) Int. of \$1 for 24 da. = \$.004; \$1 face costs \$1 — \$.004 —\$.00875 = $\$.98\frac{29}{40}$; $\$6836.75 \div .98\frac{29}{40} = \$6925.04$, *Ans.*

(9.) Bacon brought $\$.11\frac{1}{2} \times 5560 = \$639.40$; commission from each \$1 leaves \$.975; and hence draft remitted (exchange being \$.985) = $\$639.40 \times \frac{975}{985} = \$632.91$, *Ans.*

(10.) Corn brought \$4750; this, less 3%, was purchase money, \$4607.50; discount on \$1 for 63 da. = \$.0105, and proceeds, \$.9895; hence, each \$1 in face of draft costs \$.9895 + \$.02 = \$1.0095; \$4607.50 ÷ 1.0095 = \$4564.14, *Ans.*

(11.) \$20312.50 × (1 — .015) ÷ (1 — .00½) = \$20108.35, *Ans.*

(13.) Bank disc. of \$250 for 93 da. = \$3.875; proceeds = \$246.125; \$246.125 — \$244.25 = \$1.87½, gain; \$1.875 ÷ \$250 = ¾% discount, *Ans.*

(14.) (\$1011.84 — \$992) ÷ \$992 = 2%, *Ans.*

FOREIGN EXCHANGE.

Art. 321.

(2.) £625 10s. 10d. = £625.541⅔; \$4.87 × 625.541⅔ = \$3046.39—, *Ans.*

(3.) \$1 = 5.15 fr.; \$1485 ÷ 5.15 = \$288.35, *Ans.*

(5.) 5000 roubles ÷ .74 = $6756\frac{28}{37}$ roubles, *Ans.*

(6.) 1 milreis = 54 ct.; 4500 ÷ .54 = 8333.333+; 2% of it = 166.666+; hence, taking diff., in milreis, 8166.66⅔, *Ans.*

(7.) 1000 guilders = \$1000 × .40¼ = \$402.50
Disc. for 63 days = \$402.50 × .0105 = 4.226+
Proceeds, \$398.273
Brokerage to be paid, ⅛% = .499, say .50
Balance, \$397.77, *Ans.*

ARBITRATION OF EXCHANGE.

Art. 322. (1.)

\$6000 × 1.005 = \$6030, direct. N. Y. \$6000 = (\$) Gal.
Gal. \$1.00¼ = \$1 N. O. N. O. \$ .99¾ = \$1 N. Y.
\$6000 × 1.00¼ × .99¾ = \$5999.96, circular, *Ans.*
\$6030 — 5999.96 = \$30.04, gain, *Ans.*

(2.)

$\$7165.80 \div 1.00\frac{1}{4} = \$7147.93$, direct, *Ans.*

St. L. $7165.80 = ($) Balt.
Balt. $1 = $ .99¾ Hav.
Hav. $1 = $ .99⅞ N. O.
N. O. $1 = $1.00⅛ St. L.

$$\frac{\$7165.80}{.99\frac{3}{4} \times .99\frac{7}{8} \times 1.00\frac{1}{8}} = \frac{\$7165.80}{.99749844+} = \$7183.770, \text{ circ. } Ans.$$

$\$7183.770 - 7147.93 = \$35.84$, gain. *Ans.*

(3.)

$\$10000 \times 1.00\frac{1}{8} = \$10012.50$, direct.

C. $10000 = ($) L.
L. $1.00½ = $1 N. Y.
N. Y. $1 = $1.00⅕ C.

$$\frac{\$10000 \times 1.005}{1.002} = \$10029.94, \text{ circular, } Ans.$$

$\$10029.94 - 10012.50 = \$17.44$, gain, *Ans.*

(4.)

Period of discount from April 26 to June 7, 1 mon. 12 da.; and bank proceeds $= \$5284.67 \times (1 - .007) = \$5247.6773$. On each $1 in the N. Y. bill there is a commercial discount of ½ ct., and int. off for 10 da. would be additional discount $.001⅔, in all, $.006⅔, leaving each N. Y. dollar worth $.993⅓. Hence, the N. Y. bill $= \$5247.6773 \div .99\frac{1}{3} = \$5282.896$; and ¼% premium being $13.207, the amount in Cin. $= \$5282.896 + \$13.207 = \$5296.103$, *Ans.*

From May 3, to June 7, 35 da.; the interest to come off would have been a $\frac{7}{12}\%$ discount; the $\frac{1}{8}\%$ premium would reduce the discount to $\frac{11}{24}\%$; hence, the direct draft on N. O. would have brought only \$5284.67 $\times$ (1 — .00$\frac{11}{24}$) = \$5260.45 ; and \$5296.103 — \$5260.45 = \$35.65, gain, *Ans.*

(5.)

$\$\frac{10000}{.238} \times \frac{96}{4} = \$10084.033$, direct. (Art. 184.)

$\$\frac{10000}{.238}$ R. = (\$)
\$4.90 = .99$\frac{7}{8}$ £.
£. 1 = 25.38 f. $\times$ .99$\frac{7}{8}$.
5 f. = 4 R.

$$\frac{42016.807 \times 4.90 \times 5}{.99\frac{7}{8} \times 25.38 \times .99\frac{7}{8} \times 4} = \frac{1029411.7715}{101.26645} = \$10165.38$$

$$\$10165.38 - \$10084.03 = \$81.35, \textit{ Ans.}$$

EQUATION OF PAYMENTS.

Art. 326.

1. A. *to* B. Dr.

1877.			\$	ct.	When due.		Days after Sep. 15,	Prod.
May	15	To invoice at 4 mon.,	800		Sept.	15	0	0
June	1	" " " 4 mon.,	700		Oct.	1	16	11200
"	10	" " " 4 mon.,	900		"	10	25	22500
July	20	" " " 4 mon.,	600		Nov.	20	66	39600
Aug.	1	" " " 4 mon.,	500		Dec.	1	77	38500
"	15	" " " 4 mon.,	1000		"	15	91	91000
			4500					202800

4500)202800(45 da. *after* Sept. 15=Oct. 30, *Ans.*

2. Dr. *E. in acc't current with F.* Cr.

			\$	ct.	Prodc'ts from May 4.				\$	ct.	Prod. from May 4
Feb.	4	To inv., 3 mon.,	550		0	May	8	By cash.	150		600
Mar.	20	" 3 mon.,	260		12220	"	26	" remittance, June 5,	420		13440
Apr.	1	" 3 mon.,	150		8700						
"	5	" 3 mon ,	325		20150	June	3	" note, 1 mon.	340		20400
						July	1	" " 2 "	170		20400
			1285		41070				1080		54840
			1080								41070
			\$205								13770

E. owes F. \$205)13770(67 da. *before* May 4=Feb. 26, *Ans.*

3. *H. Wright to Mason & Giles.* Dr.

1876.			$	ct.	When due.		Da. after Apr. 8.	Products.
Feb.	1	To invoice, 3 mon.,	900		May	1	23	20700
"	20	" " 3 "	700		"	20	42	29400
Mar.	10	" " 3 "	600		June	10	63	37800
Apr.	8	" " cash,	500		Apr.	8	0	0
May	10	" " 3 "	900		Aug.	10	124	111600
June	15	" " 3 "	400		Sept.	15	160	64000
		Total	$4000					263500

4000)263500

66 da. *after* Apr. 8=June 13, *Ans.*

4. Dr. *A. in account current with B.* Cr.

			$	ct.	Products after Feb. 20.				$	ct.	Prod'cts after Feb. 20.
Mar.	19	To invoice cash,	900		24300	Feb.	20	By cash,	400		0
Apr.	20	" " "	800		47200	Mar.	5	" remit.			
May	10	" " "	600		47400			Mar. 15,	300		6900
June	15	" " "	700		80500	June	20	" cash,	200		24000
						July	10	" "	500		70000
			3000		199400				1400		100900
			1400		100900						
			1600		98500						

A owes B $1600)98500(62 da. *after* Feb. 20=Apr. 23, *Ans.*

5. Dr. *C. in acc't current with D.* Cr.

			$	ct.	Products after Mar. 3.				$	ct.	Products after Mar. 3.
Jan.	4	To invoice 2mon,	250		250	Mar.	10	By cash	350		2450
Feb.	3	" " 1 "	140		0	"	21	" "	200		3600
"	15	" " 2 "	450		19350	Apr.	4	" note 2 mo.	240		22320
Apr.	2	" " cash,	100		3000	May	20	" remit. May 25,	120		9960
			940		22600	June	16	" accep. 16 da.sight.	500		60500
									1410		98830
									940		22600
								Due C,	$470		76230

$\frac{76230}{470}$=162 da. *after* Mar. 3=Aug. 12, *Ans.*

(6.) The discount on $912 for 15 da. = the discount on $1 for 912 times 15 da. = 13680 da.; and the discount of $1 for 13680 da. = that on $500 for $\frac{13680}{500}$, or 27 da. nearly; hence, the $500 must be paid 27 da. after Dec. 20, which is Jan. 16, next, *Ans.*

(7.) The one was due in 64 days, the other in 34 days. By the rule, had we sought the equated time, knowing the amounts, we should have multiplied one by 64 and the other by 34, and divided the sum of the products by 375, obtaining 44; that is, 64 times one and 34 times the other = \$16500; therefore, 34 times both and 30 times one = \$16500. But 34 times both is the same as 34 times \$375, or \$12750; hence, 30 times one must be \$3750, and $\frac{1}{30}$ of \$3750, or \$125, is that one; \$250 the other, *Ans.*

(8.) Counting from Oct. 3, \$840 × 0 = 0; 400 × 94 = 37600; 200 × 63 = 12600; 37600 + 12600 = 50200, in A's favor; he still owes \$840 — \$400 — \$200 = \$240, which he should retain 50200 ÷ 240 = 209 da. after due; 209 da. after Oct. 3, is April 30th, of next year, *Ans.*

(9.) Counting from Oct. 25, \$3200 × 0 = 0; 400 × 40 = 16000; 800 × 25 = 20000; 16000 + 20000 = 36000; hence, I should have a discount on \$1 for 36000 da., for paying part, before due; \$3200 — \$400 — \$800 = \$2000; 36000 ÷ 2000 = 18 da. after Oct. 25, which is Nov. 12, *Ans.*

(10.) \$2500 — \$500 — \$500 — \$500 = \$1000; counting from Sept. 16, \$2500 × 0 = 0; 500 × 46 = 23000; 500 × 36 = 18000; 500 × 26 = 13000; 23000 + 18000 + 13000 = 54000; 54000 ÷ 1000 = 54 da. after Sept. 16, which is Nov. 9, *Ans.*

	Debts.	Terms.	Equiv.	Debts.	Terms.	Equiv.
(11.)	\$1200 ×	41 =	49200	\$1300 ×	1 =	1300
	1500 ×	72 =	108000	1300 ×	2 =	2600
	2050 ×	80 =	164000	1300 ×	3 =	3900
	1320 ×	110 =	145200	1300 ×	4 =	5200
	1730 ×	125 =	216250	1300 ×	5 =	6500
	6)\$7800		682650	1300 =	6 =	7800
	\$1300 each, *Ans.*					27300

682650 ÷ 27300 = 25 da. interval, and the notes run 25, 50, 75, 100, 125 and 150 da. respectively, *Ans.*

(12.) One drew int. for 5 mo., the other was discounted 7 mo. before due. On each \$1 of the first the gain was $\frac{5}{12}$ of 6 ct., that is, $\frac{1}{40}$ of \$1. Each \$1 of the second was bought at its true worth 7 mo. before due, that is, \$1 was bought for $\frac{1}{1.035}$ of a dollar; hence, the true discount of a dollar was $1 - \frac{1}{1.035} = \frac{7}{207}$ of a dollar. Hence, since the interest drawn and the discount taken were equal, $\frac{1}{40}$ of the first was equal to $\frac{7}{207}$ of the other; that is, one was $\frac{280}{207}$ of the other; *both*, then, must have been $\frac{487}{207}$ of that other, and since both made \$487, the first was \$280, and the other \$207. Now, by the common rule, equating, we should have

$$\begin{array}{lll} 280 \times 1 = & 280. & \\ 207 \times 2 = & \underline{414} & \\ \overline{487} &)694(& 1.42505 \text{ yr.} = 1 \text{ yr. } 5 \text{ mo. } 3 \text{ da.} \end{array}$$

Hence, the difference is 3 days, *Ans.*

SETTLEMENT OF ACCOUNTS.

Art. 329.

(1.)

HENRY HAMMOND.

DR. *Focal Date, Apr. 4.* CR.

1876.		Sums.	Products.
Apr.	4	\$900 × 0 =	0
June	1	400 × 58 =	23200
May	18	701 × 44 =	30844
July	7	600 × 94 =	56400
Apr.	8	500 × 4 =	2000
June	9	400 × 66 =	26400
July	1	101 × 88 =	8888
		3602)	147732(41 da. after Apr. 4, = May 15, *Ans.*

Then the exact sum (\$3601.80) held beyond May 15, to July 7, draws interest for 53 da., and the am't = \$3601.80 × ($1.008\frac{5}{6}$) = \$3633.62, *Ans.*

\$3601.80 due May 15, settled 15 days earlier, amounts to \$3601.80 × .9975 = \$3592.80, *Ans.*

(2.)

WILLIAM SMITH.

DR. *Focal Date, Jan. 1.* CR.

1876.	Due.	da.	$	prod.
Jan. 1,		0 ×	800 =	0
" 16,	Feb. 15,	45 ×	180 =	8100
Feb. 14,	Apr. 14.	104 ×	401 =	41704
Mar. 25,	Mar 25,	84 ×	500 =	42000
Apr. 1,	Apr. 1,	91 ×	800 =	72800
May 7,	May 7,	127 ×	600 =	76200
" 21,	July 20.	201 ×	700 =	140700
June 10,	June 10,	161 ×	200 =	32200
" 15,	Sept. 13,	256 ×	2000 =	512000
July 12,	July 12,	193 ×	500 =	96500
Aug. 4,	Oct. 4,	277 ×	1000 =	277000
			$7681	1299204
			5361	908034
			$2320	391170

1876.	Due.	da.	$	prod.
Jan. 10,	Jan. 10,	9 ×	400 =	3600
" 28,	" 28,	27 ×	200 =	5400
Feb. 15,	Apr. 15,	105 ×	180 =	18900
" 28,	Feb. 28,	58 ×	100 =	5800
Mar. 30,	June 28,	179 ×	450 =	80550
Apr. 14,	Apr. 14,	104 ×	401 =	41704
May 1,	May 1,	121 ×	500 =	60500
" 15,	June 15,	166 ×	680 =	112880
June 16,	" 16,	167 ×	300 =	50100
July 19,	July 19,	200 ×	700 =	140000
Aug 10,	Aug. 30,	242 ×	200 =	48400
Sept. 1,	Sept. 1,	244 ×	150 =	36600
Oct. 3,	Oct. 3,	276 ×	1100 =	303600
			5361	908034

391170 ÷ 2320 = 168.6; hence, 169 da. after Jan. 1, 1876, which is June 18, 1876, *Ans.* Int. of $2320.10 from June 18, to Oct. 4, (108 da.) 10% = $69.60 am't = $2389.70, *Ans.*

(3.)

GEORGE CUMMINGS.

1876. DR. *Focal Date, Jan. 1.* CR.

Due.	da.	$	prod.
Jan. 2,	1 ×	300 =	300
" 21,	20 ×	194 =	3880
Mar. 4,	63 ×	150 =	9450
		$644	13630

Due.	da.	$	prod.
Jan. 1,	0 ×	583 =	0
Mar. 3,	62 ×	40 =	2480
Apr. 30,	120 ×	130 =	15600
		$753	18080
		644	13630
		109	4450

4450 ÷ 109 = 40.8+; hence, 41 da. after Jan. 1, or, Feb. 11th, *Ans.* $109 on int. to Mar. 31 49 da., amounts to $109 × 1.01361 = $110.48, *Ans.*

(4.)

DR. *A. L. Morris in acc't with T. J. Fisher & Co.* CR.

$		da.	int.
350.	6%	168	$9.800
275.	"	159	7.287
100.	"	125	2.083
400.	"	60	4.00
108.25	"	7	.126
			$23.296

$	da.		int.
813.64	181		$24.544
120.	146		2.92
500.	105		8.75
85.	30		.425
			$36.639
			23.296
		Int. bal.	$13.34
		Debt.	274.89
		Due,	$298.23

(Rule on p. 293, Higher Arith.)

(5.)

Wm. White in acc't with Beach & Berry.

DR. (Rule on p. 293.) CR.

$		da.	int.	$	da.	int.
212.50	10%	102	$6.020	1102.50	104	$31.850
66.	"	84	1.540	50.	98	1.361
235.	"	66	4.308	95.	89	2.348
300.	"	48	4.000	168.75	64	3.000
110.	"	37	1.130	32.	24	.213
46.40	"	31	.399	79.90	9	.199
454.25	"	15	1.892			$38.971
			$19.289			19.289
					Int. bal.	$19.682
					Debt,	104.
					Due,	$113,68, *Ans.*

ACCOUNT SALES.

Art. 331. (1.)

Maynard's Consignment.

(Focal date, Aug. 10.)

SALES. (1st Calculation.)

Due.		*Days.*		*Dol.*		*Prod.*
Aug.	12.	2	×	50.60	=	101.20
"	14.	4	×	800.	=	3200.
Sept.	23.	44	×	850.	=	37400.
Aug.	29.	19	×	210.	=	3990.
"	30.	20	×	4900.	=	98000.
Sept.	20.	41	×	1400.	=	57400.
				$8210.60		200091.20

CHARGES. (2d Calculation.)

Due.		*Days.*		*Dol.*		*Prod.*
Aug.	10.	0	×	75.	=	0
"	31.	21	×	10.	=	210.
"	31.	21	×	10.26	=	215.46
Sept.	3.*	24	×	205.25	=	4926.50
				300.51		5351.96
				8210.60		200091.20
				7910.		194739.24

Balance of the two.—194739.24 ÷ 7910 = 24.6; hence the equated time is 25 da. past Aug. 10, which is Sept. 4, 1876, *Ans.*

---

* 200091.20 ÷ 8210 = 24.3; hence, take 24 da. *after* Aug. 10, *for date of commission*, on other side; *i. e.*, Sept. 3.

(2.)

ACCOUNT. *W. Thomas sold on acc't of B. F. Jonas.*

CHARGES.			SALES.		
1876.		$	1876.		$
July 6.	To Freight paid	150.	July 8.	2000 bu. wheat........	2112.50
" 11.	" Storage	6.	" 11.	300 " on 20 da.....	362.50
" 11.	" Drayage	5.			$2475.
" 11.	" Insurance	4.			
" 11.	" Commission	61.87			
" 11.	" Loss and Gain.....	11.57			
		238.44			

CALCULATION FROM FOCAL DATE, JULY 6.

Due.	*Day.*	*Dol.*	*Prod.*	*Due.*	*Day.*	*Dol.*	*Prod.*
July 6.	0 ×	150.	= 0	July 8.	2 ×	2112.50	= 4225.00
" 11.*	5 ×	88.44	= 442.20	" 31.	25 ×	362.50	= 9062.50
		$238.44				2475.	13287.50
						238.44	442.20
				Net proceeds,		$2236.56	12845.30

Since on the side ot the sales we find 13287.50 contains 2475, *five* times, 5 da. *after* July 6, or July 11, is the date the *commission* should have, in the calculation. Its *entry* and *time due* are of the *same* date, (which was *not* the case in the preceding example).

12845.30 ÷ 2236.56 = 5.7; hence, take 6 days after July 6, or July 12, 1876, *Ans.*

(3.)

Mdse. Co. "B."

SALES. CHARGES.

CALCULATION FROM FOCAL DATE, JULY 15.

Due.	*Days.*	*Dol.*	*Prod.*	*Due.*	*Days.*	*Dol.*	*Prod.*
Aug. 7.	23 ×	120	= 2760	July 15.	0 ×	37.00	= 0
18.	34 ×	60	= 2040	" 30.	15 ×	4.50	= 67.50
July 30.	15 ×	55	= 825	Aug. 8.*	24 ×	5.29	= 126.96
		$235	5625	July 30.	15 ×	62.73	= 940.95
						$109.52	1135.41

Equated thus: $235 — $105.59 = $129.41; 5625. — 1135.41 = 4489.59; and 4489.59 ÷ 129.41 = 34.7; hence, take 35 da. after July 15; *i. e.*, Aug. 19, *Ans.*

---

*On first side calculate the *date* to be given to *commission* on the other side, thus: 5625 ÷ 235 = 23.9; hence, take 24 da. after July 15, *i. e.*, Aug. 8.

Carnes's \$62.73, on interest from August 19 to January 1 (4 mo. 12 da.), 10%, amounts to \$62.73 + \$2.30 = \$65.03, net proceeds, *Ans.*

STORAGE ACCOUNTS.

Art. 332.

(1.)

RECEIVED. DELIVERED.

1876.		*Bbl.*	*Bal. on hand.*	*Days.*	*Products.*	1876.		*Bbl.*
January	2	200	200	3	600			
"	5	150	350	2	700			
"	7	30	380	3	1140			
"	10	120	500					
			390	3	1170	January	10	110
			300	1	300	"	13	90
"	14	80	380	3	1140			
"	17	150	530					
			510	3	1530	"	17	20
"	20	75	585					
			470	4	1880	"	20	115
"	24	60	530	1	530			
			390	2	780	"	25	140
			318	1	318	"	27	72
"	28	200	518	2	1036			
	bbl	on hand	– 418	1	418	"	30	100
				30)	11542			
					385	av. no. of	bbl.	stored.
					.05			
					\$19.25	Storage.		

(2.)

RECEIVED. DELIVERED.

		Bbl.	*Bal. on hand.*	*Days.*	*Products.*			*Bbl.*
January	31	418	418	4	1672			
February	4	250	668	1	668			
			568	4	2272	February	5	100
"	9	120	688	1	688			
			608	2	1216	"	10	80
"	12	100	708					
			488	2	976	"	12	220
			348	2	696	"	14	140
"	16	30	378	2	756			
			288	2	576	"	18	90
"	20	bbl. on	hand, 0	0	0	"	20	288
				30)	9520			
					317	av. no. of	bbl.	stored.
					.05			
					\$15.85	Storage.		

COMPOUND INTEREST.

CASE I.

Art. 334.

(1.)

\$3850
5
192.50
3850
4042.50
5
202.125
4042.50
4244.625
5
212.231
4244.625
4456.856
5
222.843
4456.856
4679.699
5
233.985

6 mo. = $\frac{1}{2}$ | 116.993
1 mo. = $\frac{1}{6}$ | 19.499
15 da. = $\frac{1}{2}$ | 9.749
1 da. $\frac{1}{15}$ | .650

146.891
4679.699
Amount \$4826.59, *Ans.*
3850.
Interest \$976.59, *Ans.*

(2.) \$13062.50
2
261.25
13062.50
13323.75
2% = 266.475
13590.225
2% = 271.805
13862.03
2% = 277.241
14139.271
2% = 282.785
14422.056
2% = 288.441
14710.497
2% = 294.210
15004.707
1 mo. 12 da. = 140.044
Amount = 15144.751 [*Ans.*
15144.75 — 13062.50 = \$2082.25

(3.) \$1000
1.05
1050
1.05
1102.50
1.05
Amt. for 3 yr. 1157.625
1.05×1.05×1.05 = 1.157625
Ans. Prod. \$1340.10

(4.)

	\$2000.
3% =	60.
	2060.
3% =	61.80
	2121.80
3% =	63.654
	2185.454
3% =	65.564
	2251.018
$.029\frac{1}{2}$ =	66.405
$1.149\frac{1}{2}$)	2317.423
	\$2016.03, *Ans.*

(5.)

	Annual.	*Compound.*
	\$5000.	\$5000.
.06 × 6 =	.36	300.
	\$1800.	5300.
\$300 × .9 =	270.	318.
	\$2070.	5618.
		337.08
		5955.08
		357.3048
		6312.3848
		378.7431
		6691.1279
		401.4677
		7092.5956
		5000.
		2092.5956 (last digit struck through)
		2070.
		Ans. \$22.596

(6.)

	\$1000.
1st yr. int.	210.
	1210.
2d yr. int.	254.10
	1464.10
	$1.10\frac{1}{2}$
	73205
	1610510
	\$1617.83, *Ans.*

CALCULATION BY TABLES.

Art. 335.

(1.) \$750 × 2.6927728 = \$2019.58, *Ans.*

(2.) \$5428 × 5.0031885 = \$27157.31, *Ans.*

(3.) 28 intervals at 4%; \$1800 × 2.99870332 = \$5397.67; \$5397.67 — \$1800 = \$3597.67, *Ans.*

(4.) 42 intervals at $3\frac{1}{2}$%; \$1000 × 4.24125799 = \$4241.26, *Ans.*

(5.) 38 intervals at $4\frac{1}{2}\%$; \$9401.50 × 5.32621921 = \$50074.45; interest of \$50074.45 for 4 mo. at 9% = \$1502.23; \$50074.45 + \$1502.23 = \$51576.68 compound amount, *Ans.*

(6.) 100 yr. = 50 yr. + 50 yr.; \$1000 × 117.3908529 × 117.3908529 = \$13780612.34, *Ans.*

(7.) 60 intervals at 2%; 60 = 55 + 5; \$3600 × 2.97173067 × 1.1040808 = \$11811.71, am't; \$11811.71 — \$3600 = \$8211.71 int., *Ans.*

(8.) 80 intervals at $2\frac{1}{2}\%$; 80 = 50 + 30; \$4000 × 3.43710872 × 2.09756758 = \$28838.27; \$28838.27 — \$4000 = \$24838.27, *Ans.*

(9.) There are 111 intervals at 3%; 111 = 55 + 55 + 1; \$1200 × 5.08214859 × 5.08214859 × 1.03 = \$31923.70; amount of \$31923.70 for 2 mo. 4 da. at 12% = \$32604.74; \$32604.74 — \$1200 = \$31404.74, *Ans.*

CASE II.

Art. 336.

(1.) \$12000 ÷ \$8000 = 1.5; in table, this is between 6 and 7 yr.; 1.50 — 1.4185191 = .0814809; 1.5036303 — 1.4185191 = .0851112; $\frac{.0814809}{.0851112}$ yr. = 11 mo. 15 da.

Ans. 6 yr. 11 mo. 15 da.

(2.) \$5200 + \$1308 = \$6508; \$6508 ÷ \$5200 = 1.25153846; this is between 7 and 8 intervals at 3%; 1.25153846 — 1.22987387 = .02166459; 1.26677008 — 1.22987387 = .03689621; $\frac{.02166459}{.03689621}$ of 6 mo. = 3 mo. 16 da.; $\frac{7}{2}$ yr. + 3 mo. 16 da. = 3 yr. 9 mo. 16 da., *Ans.*

(3.) \$18000 ÷ \$12500 = 1.44; this is over 14 intervals at $2\frac{1}{2}\%$; 1.44 — 1.41297382 = .02702618; 1.44829817 — 1.41297382 = .03532435; $\frac{.02702618}{.03532435}$ of 3 mo. = 2 mo. 9 da.; $\frac{14}{4}$ yr. + 2 mo. 9 da. = 3 yr. 8 mo. 9 da., *Ans.*

(4.) \$1 + \$1 = \$2; \$2 ÷ \$1 = 2.; in table at 6%, 2. is between 11 and 12 yr.; 2 — 1.8982986 = .1017014; 2.0121965 — 1.8982986 = .1138979; $\frac{.1017014}{.1138979}$ yr. = 10 mo. 21 da.; at 8%, 2. is over 9 yr.; 2.— 1.9990046 = .0009954; 2.1589250 — 1.9990046 = .1599204; $\frac{.0009954}{.1599204}$ yr. = 2 da.; at 10%, 2. is over 7 yr.; 2. — 1.9487171 = .0512829; 2.1435888 — 1.9487171 = .1948717; $\frac{.0512829}{.1948717}$ yr. = 3 mo. 5 da.

Ans. 11 yr. 10 mo. 21 da.; 9 yr. 2 da.; 7 yr. 3 mo. 5 da.

(5.) \$22576.15 ÷ \$9862.50 = 2.28909; this is over 14 intervals at 6%; 2.28909 — 2.260904 = .028186; 2.3965582 — 2.260904 = .1356542; $\frac{.028186}{.1356542}$ of 6 mo. = 1 mo. 7 da.; $\frac{14}{2}$ yr. + 1 mo. 7 da. = 7 yr. 1 mo. 7 da., *Ans.*

Art. 337.

Case III.

(1.) \$1593.85 ÷ \$1000 = 1.59385, which is found in the table under 6%. *Ans.* 6%.

(2.) \$6332.51 + \$3600 = \$9932.51; \$9932.51 ÷ \$3600 = 2.75903, which is in the table, under 7%. *Ans.* 7%.

(3.) \$48049.58 ÷ \$13200 = 3.6401197; amount of 3.5556727 for 5 mo. 21 da. at 5% = 3.6401199. *Ans.* 5%.

(4.) \$13276.03 ÷ \$2813.50 = 4.71868847; amount of 4.66734781 for 1 mo. 14 da. at 4½% = 4.71868864; 4½% semi-annually = 9% per annum, *Ans.*

(5.) There are 47 intervals; \$17198.67 + \$7652.18 = \$24850.85; \$24850.85 ÷ \$7652.18 = 3.24755168; amount of 3.19169713 for 2 mo. 3 da. at 2½% a quarter = 3.2475518; 2½% quarterly = 10% per annum, *Ans.*

(6.) $1+1=2$; $2\div1=2$; looking in the tables for 2 opposite 10, 15, and 20 yr., we find for 10 yr. between 7 and 8%, *Ans.*; for 15 yr. not quite 5%, *Ans.*; for 20 yr. a little over $3\frac{1}{2}$%, *Ans.*

Case IV.

Art. 338.

(1.) Interest of \$1 for 25 yr. at 6% = \$3.2918707; \$52669.93 ÷ 3.2918707 = \$16000, *Ans.*

(2.) Int. of \$1 for 6 yr. 2 mo. at 7%, payable semi-annually = .5287; \$1625.75 ÷ .5287 = \$3075, *Ans.*

(3.) Int. of \$1 for 3 yr. 6 mo. 9 da., at 10%, payable quarterly = \$.416506; \$3598.61 ÷ .416506 = \$8640, *Ans.*

(4.) Int. of \$1, in 19 periods at 4% = \$1.10684918; \$31005.76 ÷ 1.10684918 = \$28012.63, *Ans.*

(5.) Amount of \$1 in 7 yr. at 4% = \$1.31593178; \$27062.85 ÷ 1.31593178 = \$20565.54, *Ans.*

(6.) Amount of \$1 in 5 yr. 9 mo. at 6%, payable semi-annually = \$1.40499738; \$14625.70 ÷ 1.40499738 = \$10409.77, *Ans.*

(7.) Amount of \$1 in 12 yr. 8 mo. 25 da. at 5% = \$1.8619538; \$8767.78 ÷ 1.8619538 = \$4708.91, P. worth; \$8767.78 — \$4708.91 = \$4058.87, *Ans.*

ANNUITIES.

Case I.

Art. 340.

(1.) \$300 ÷ .06 = \$5000, *Ans.*

(2.) \$756.40 ÷ .08 = \$9455, *Ans.*

(3.) \$15642.90 ÷ .07 = \$223470, *Ans.*

(4.) Interest at 5% on \$800 for 6 mo. = \$20; \$1620 ÷ .05 = \$32400, *Ans.*

(5.) Interest of \$625 for 3 mo. at 6% = \$9.37½; \$1250 + \$9.37½ = \$1259.37½, value in ½ yr.; \$1259.37½ ÷ .03 = \$41979.16⅔, *Ans.* Int. of \$625 for 9 mo., 6 mo., 3 mo. = \$28.12½, \$18.75, and \$9.37½; hence \$2500 a year, payable quarterly = \$2500 + \$28.12½ + \$18.75 + \$9.37½ = \$2556.25 yearly; \$2556.25 ÷ .06 = \$42604.16⅔, *Ans.*

The interest is 1½% per quarter; \$625 ÷ .01½ = \$41666.66⅔, *Ans.*

Art. 341.

CASE II.

(1.) Initial value of \$780 a year, = \$780 ÷ .05 = \$15600; present value = \$15600 ÷ 1.7958563 = \$8686.66, *Ans.*

(2.) Initial value of \$160 a year = \$160 ÷ .07 = \$2285.714; amount of \$1 for 3 yr. 4 mo. at 7% = \$1.2536273; \$2285.714 ÷ 1.2536273 = \$1823.28, *Ans.*

(3.) Initial value of \$540 a year = \$540 ÷ .06 = \$9000; present value = \$9000 ÷ 1.7908477 = \$5025.55, *Ans.*

(4.) 8%, payable semi-annually = .0816 per year; initial value of \$325 a year = \$325 ÷ .0816 = \$3982.843; present value = \$3982.843 ÷ 1.48024428 = \$2690.67, *Ans.*

(5.) Interest on \$250 for 3 mo. at 10% = \$6.25; the value, semi-annually, is \$500 + \$6.25 = \$506.25, of which the initial value is \$506.25 ÷ .05 = \$10125. The present value of \$10125, due in 9 yr. 10 mo. 18 da., at 10%, payable semi-annually, is \$10125 ÷ 2.6238167 = \$3858.88, *Ans.*

Art. 342.

CASE III.

(1.)

Pres. val. of perpetuity of \$125, deferred 12 yr. = \$792.88
Pres. val. of perpetuity of \$125, deferred 24 yr. = 352.05
Ans. \$440.83

(2.)

Pres. value of immediate perpetuity of \$400 = \$5000.00
Pres. val. of perpetuity of \$400, deferred $15\frac{1}{2}$ yr. = 1515.59
Ans. \$3484.41

(3.)

6%, payable semi-annually, = .0609 per year; \$826.50 ÷ .0609 = \$13571.429 initial value of perpetuity; present value of such perpetuity, deferred 3 yr., = \$13571.429 ÷ 1.1940523 = \$11365.86; present value of such perpetuity, deferred $16\frac{3}{4}$ yr., = \$13571.429 ÷ 2.69212027 = \$5041.17; \$11365.86 — \$5041.17 = \$6324.69, *Ans.*

(4.)

Pres. val. of perpetuity of \$60, deferred 12 yr. = \$786.22
Pres. val. of perpetuity of \$60, deferred 21 yr. = 529.05
Ans. \$257.17

(5.)

Int. on \$120 for 3 mo. at 8% = \$2.40; \$240 + \$2.40 = \$242.40, value half-yearly; hence the lease may be considered an immediate annuity of \$242.40 per half year, and running 8 yr. 9 mo.;

Present value of immediate perpetuity, . . = \$6060.
Present value of perpetuity, deferred $8\frac{3}{4}$ yr., = 3050.04
Present value of lease, ... \$3009.96

\$3009.96 — \$2500 = \$509.96 loss, *Ans.*

CASE IV.

Art. 343.

(1.) Initial value of a perpetuity of \$300 at 5% = \$6000; compound interest of \$6000 for 18 yr. = \$6000 × 1.4066192 = \$8439.72, *Ans.*

(2.) Initial value of a perpetuity of \$25 at 10% = \$250; compound interest of \$250 for 40 yr. = \$250 × 44.2592556 = \$11064.81, *Ans.*

(3.) Initial value of a perpetuity of \$75 at 6% = \$1250; compound interest of \$1250 for 9 yr. = \$1250 × .689479 = \$861.85, *Ans.*

(4.) Initial value of a perpetuity of \$5 at 9% = \$555$\frac{5}{9}$; compound int. of \$555$\frac{5}{9}$ for 50 yr. = \$555$\frac{5}{9}$ × 73.35752 = \$4075.42, *Ans.*

(5.) The annuity may be considered as the yearly interest of some principal at annual interest. Each \$1 in 14 yr. draws \$.06 × 14 = \$.84; and one annual interest of 6 ct., for 91 intervals (Art. 304), yields \$.06 × .06 × 91 = \$.3276; in all, the interest yielded (or, the annuity's amount for each \$1 in the principal which yields it) = 84 ct. + \$.3276 = \$1.1676. Hence the principal yielding such an interest is \$116.76 ÷ 1.1676 = \$100; and, therefore, the annuity itself \$6. Amount of an annuity of \$6, compound interest 6%, 14 yr., find thus:

Initial value, \$100; compound interest of \$100 for 14 yr. = \$126.0904. \$126.0904 — \$116.76 = \$9.33, *Ans.*

(6.) There will be 13 deposits, and the amount is that of an annuity of \$35 running 13 yr.; initial value = \$35 ÷ .10 = \$350; compound int. of \$350 for 13 yr. = \$2.4522712 × 350 = \$858.29, *Ans.*

CASE V.

Art. 345.

(1.) Present value of \$1 a year for 17 years at 7% = \$9.763223; final value of the same = \$9.763223 × 3.1588152 = \$30.840217; \$15000 ÷ 30.840217 = \$486.38, *Ans.*

(2.) \$1 per year for 16 years, commencing now, is worth \$10.105895; commencing 4 years hence, it is worth \$10.105895 ÷ 1.262477 = \$8.0048; \$4800 ÷ 8.0048 = \$599.64, *Ans.*

CASE VI.

Art. 346.

(1.) \$1000000 ÷ \$80000 = 12.5; this falls between 23 yr. and 24 yr., in the table, under 6%;

Comp. am't of \$1000000 for 23 yr. at 6% =	\$3819749.70
Final value of \$80000 per yr. for 23 yr. =	3759666.27
Bal. due at end of 23 yr.	\$60083.43, *Ans.*

(2.) \$30000000 ÷ \$2000000 = 15; this falls between 28 and 29 yr.;

Comp. am't of \$30000000 for 28 yr. at 5% =	\$117603873.
Final value of \$2000000 per yr. for 28 yr. =	116805164.
Bal. due at end of 28 yr.	\$798709, *Ans.*

(3.) \$22000 ÷ \$2500 = 8.8; this falls between 14 and 15 yr.;

Comp. am't of \$22000 for 14 yr. at 7% =	\$56727.75.
Final value of \$2500 per yr. for 14 yr. =	\$56376.22.
Bal. due at end of 14 yr.	\$351.53, *Ans.*

(4.) In the former case \$689.61 is the principal at the end of 15 yr., or a balance which can only draw *simple* interest for a fraction of an interval. But the *amount* of that \$689.61 must be exactly equal to such a fraction of \$1000 as the part of a year is of the whole. The interest of \$689.61 for that fraction of a year must be 6% of \$689.61 × that fraction; that is, 689.61 with 41.3766 times the fraction is the same as 1000 times the fraction; hence, 689.61 *alone* must be only 958.6234 times the fraction, which must therefore be $\frac{689.61}{958.6234}$ of a year, = .71937 of a yr. = 8 mo. 18.67 da.; hence whole time = 15 yr. 8 mo. 19 da., *Ans.*

(5.) Quotient = 20, which, by table, takes more than 41 yr., *Ans.* In that time,

Comp. am't of \$2000000000, 4%	= \$9986122900.
Final value of annuity of \$100000000 =	9982653625.
Unpaid,	\$3469275, *Ans.*

Art. 347.

Case VII.

(1.) \$9000 ÷ \$750 = 12, which is about half way between 12.46 and 11.47, the values of \$1 for 20 yr., at 5 and 6%. *Ans.* About $5\frac{1}{2}$%.

(2.) \$650 ÷ \$80 = 8.125, which falls between 8 and 10%, and near 8. *Ans.* 8%+.

CONTINGENT ANNUITIES.

Art. 351.

Case I.

(1.) \$650 × 5.162 = \$3355.30, *Ans.*

(2.) 6% of \$25000 = \$1500; \$1500 × 9.524 = \$14286, life-estate, *Ans*; \$25000 — \$14286 = \$10714, rev., *Ans.*

(3.) $\frac{1}{3}$ of \$46250 = \$15416.67, 6% on which = \$925; \$925 × 13.769 = \$12736.33, dower, *Ans.*

\$15416.67 — \$12736.33 = \$2680.34, rev., *Ans.*

CASE II.

Art. 352.

(1.) \$500 ÷ 13.368 = \$37.40, *Ans.*

(2.) \$1200 ÷ 12.957 = \$92.61, *Ans.*

(3.) \$840 ÷ 8.153 = \$103.03, *Ans.*

CASE III.

Art. 353.

(1.) Pres. value of perpetuity of \$1, at 5%, .	\$20.000
Pres. value of annuity of \$1, age 47 yr.,	12.301
Pres. value of reversion of \$1, ...	\$7.699

\$500 × 7.699 = \$3849.50, *Ans.*

(2.) Pres. value of immediate annuity of \$1 for 30 yr., ...	\$13.764831
Pres. value of annuity of \$1, age 38 yr.,	12.239
Pres. value of reversion of \$1, ...	\$1.525831

\$165 × 1.525831 = \$251.76, *Ans.*

(3.) Pres. value of \$1 lease, for 40 yr., . . .	\$13.331709
Pres. value of \$1 life-estate, age 62 yr.,	7.403
Pres. value of reversion of \$1, . . .	\$5.928709

\$1600 × 5.928709 = \$9485.93, *Ans.*

LIFE INSURANCE.

Art. 355.

(1.) Table number for 40 yr., \$31.30; \$31.30 × 5 = \$156.50, *Ans.*

(2.) From 40 to 53, inclusive, 14 yr.; \$156.50 × 14 = \$2191, *Ans.*

(3.) \$2191 in all; one payment on interest 13 yr., 1

for 12, and so on to 1 for 1 yr.; all equivalent to interest on one for 91 yr.; hence,

Payments, ...	$2191.
Int. on $156.50 for 91 yr., 6%, . . .	854.49
	$3045.49, *Ans.*

(4.) Endowment policy cost per year $112.68; life policy $47.18; difference on each $1000, $65.50; $65.50 × 120 = $7860, *Ans.*

(5.) End'm't premiums $1042.90 × 2 = $2085.80
Int. on one for 2 yr., 6%, . . . 125.148
$2210.948

Life pol. cost $214.80×2=$429.60
Int. for 2 yr. on *one*, . . . 25.776— $455.376
Profit, $1755.57, *Ans.*

$10000 — $2210.948 = $7789.052, *Ans.*

(6.) On each $1000, per yr. $31.30; ten of them = $313; annual interest for 9 yr., = $31.30 × .06 × 45 = $84.51; $313 + $84.51, or $397.51, on *one* thousand, hence the $3975.10 must have been on $10000, *Ans.*

(7.) 15 premiums, of $1 each, and the annual interest amount to $15 + interest of $1 for 120 yr., and = $22.20; on $1000 the amount is given $1542.648; and $1542.648 ÷ 22.20 = $69.49, corresponding in table to 40 yr., *Ans.*

(8.) Annual payment for $8000 = $19.80 × 8 = $158.40; this, in $8000 is contained more than 50 and less than 51 times; hence 51st is the payment making the excess; 20 yr. + 51 yr. = 71 yr., *Ans.*

(9.) Annual payment $9.919 × 300 = $2975.70; 3 payments and interest for 3½ yr. on *one*, amounts to $2975.70 × 3.21 = $9551.997; $30000 — $9552 = $20448, *Ans.*

(10.) Amount paid = \$11635.80; on \$1000, \$1057.80; the table shows 10, 15, 20; dividing by the sums opposite 42 yr., at the third we have exactly 20 yr., *Ans.*

PARTNERSHIP.

Case I.

Art. 357.

(1.) \$3600 — \$1500 = \$2100 gain; \$2500 + \$1875 = \$4375; $\frac{2500}{4375}$ of \$2100 = \$1200, A, *Ans.*; $\frac{1875}{4375}$ of \$2100 = \$900, B, *Ans.*

(2.) \$7000 — \$800 — \$1000 = \$5200 to be divided; 50 + 64 + 16 = 130;

$\frac{50}{130}$ of \$5200 = \$2000, A's gain, *Ans.*
$\frac{64}{130}$ of \$5200 = \$2560, B's gain, *Ans.*
$\frac{16}{130}$ of \$5200 = \$640 + \$1000 = \$1640, C's income, *Ans.*

(3.) \$24000 + \$28000 + \$32000 = \$84000; $\frac{1}{6}$ of \$84000 = \$14000; \$84000 — \$14000 = \$70000; $\frac{3}{5}$ of \$70000 = \$42000; \$70000 + \$42000 = \$112000; \$112000 — \$84000 — \$8000 = \$20000 gain; $\frac{24}{84}$ of \$20000 = \$5714.28$\frac{4}{7}$, A's gain, *Ans.*; $\frac{28}{84}$ of \$20000 = \$6666.66$\frac{2}{3}$, B's gain, *Ans.*; $\frac{32}{84}$ of \$20000 = \$7619.04$\frac{16}{21}$, C's gain, *Ans.*

(4.) C receives $\frac{1120}{3920} = \frac{2}{7}$ of the gain; he, therefore, contributed $\frac{2}{7}$ of the capital, and A and B $\frac{5}{7}$, which, consequently = \$12960; this is $\frac{5}{7}$ of \$18144, total capital; $\frac{2}{7}$ of \$18144 = \$5184, C's stock, *Ans.*; $\frac{5760}{18144}$ of \$3920 = \$1244.44$\frac{4}{9}$, A's gain, *Ans.*; $\frac{7200}{18144}$ of \$3920 = \$1555.55$\frac{5}{9}$, B's gain, *Ans.*

(5.) \$8000 + \$12800 — \$15200 = \$5600; A and B together have \$5600 more than C, and gain \$1638 more; hence the gain is $\frac{1638}{5600} = \frac{117}{400}$ of the stock; $\frac{117}{400}$ of \$8000 = \$2340, A's, *Ans.*; $\frac{117}{400}$ of \$12800 = \$3744, B's, *Ans.*; $\frac{117}{400}$ of \$15200 = \$4446, C's, *Ans.*

(6.) \$20000 + \$16000 + \$12000 = \$48000 on quitting. $\frac{20}{48}$ of \$27000 = \$11250, A's, *Ans.*; $\frac{16}{48}$ of \$27000 = \$9000, B's, *Ans.*; $\frac{12}{48}$ of \$27000 = \$6750, C's, *Ans.*

(7.) Adding the given sums, we have \$5400, the equivalent of 3 times A's, B's, C's, D's gains altogether; and, therefore, the whole gain being \$1800, the gain of the first three being \$1150, the gain of the fourth man, D, is \$1800 — \$1150 = \$650; and if this be 30% of his stock, that stock must be \$650 ÷ .30 = \$2166$\frac{2}{3}$, *Ans.*; in like manner, $\frac{10}{3}$ of (\$1800 — \$1650) = \$500, C's, *Ans.*; $\frac{10}{3}$ of (\$1800 — \$1000) = \$2666$\frac{2}{3}$, A's, *Ans.*; and $\frac{10}{3}$ of (\$1800 — \$1600) = \$666$\frac{2}{3}$, B's, *Ans.*

PARTNERSHIP WITH TIME.

CASE II.

Art. 358.

(1.) A has \$6000 × 12 = \$72000; B, \$10000 × 6 = \$60000; 72 + 60 = 132; $\frac{72}{132}$ of \$3300 = \$1800, A's gain, *Ans.*; $\frac{60}{132}$ of \$3300 = \$1500, B's gain, *Ans.*

(2.) $\frac{2}{3}$ of 4 = 2$\frac{2}{3}$; $\frac{3}{4}$ of 5 = 3$\frac{3}{4}$; A contributes 4 parts for 3 mo., and 1$\frac{1}{3}$ parts for 9 mo., in all 24 for 1 mo.; B contributes 5 parts for 3 mo., and 1$\frac{1}{4}$ parts for 9 mo., in all 26$\frac{1}{4}$ for 1 mo.; stock-equivalents = 24 and 26$\frac{1}{4}$, or 96 and 105, or 32 and 35; 32 + 35 = 67; A gets $\frac{32}{67}$ of \$1675 = \$800, *Ans.*; B, $\frac{35}{67}$ of \$1675 = \$875, *Ans.*

(3.) A contributes $\frac{1}{2}$ for 4 mo., and $\frac{1}{4}$ for 9 mo., in all 4$\frac{1}{4}$ parts for 1 mo.; B, $\frac{1}{3}$ for 13 mo. = 4$\frac{1}{3}$ for 1 mo.; C, $\frac{1}{4}$ for 13 mo. = 3$\frac{1}{4}$ for 1 mo.; stock-equivalents = 4$\frac{1}{4}$, 4$\frac{1}{3}$, 3$\frac{1}{4}$, or 51, 52, 39; 51 + 52 + 39 = 142; A gains $\frac{51}{142}$ of \$1988 = \$714, *Ans.*; B, $\frac{52}{142}$ of \$1988 = \$728, *Ans.*; C, $\frac{39}{142}$ of \$1988 = \$546, *Ans.*

(4.) A has $2500 × 18 = $45000; B, $1500 × 18 = $27000; C, $5000 × 9 = $45000; 45 + 27 + 45 = 117; A gains $\frac{45}{117}$ of $3250 = $1250, *Ans.*; B gains $\frac{27}{117}$ of $3250 = $750, *Ans.*; C gains $\frac{45}{117}$ of $3250 = $1250, *Ans.*

(5.) A contributes $1000 for 3 mo., $600 for 3 mo., $200 for 6 mo.; B, $1000 for 3 mo., $1400 for 3 mo., $1800 for 6 mo.; 1000 × 3 = 3000; 600 × 3 = 1800; 200 × 6 = 1200; sum = 6000; 1000 × 3 = 3000; 1400 × 3 = 4200; 1800 × 6 = 10800; sum = 18000; 6000 + 18000 = 24000; $\frac{6}{24}$ of $800 = $200, A, *Ans.*; $\frac{18}{24}$ of $800 = $600, B, *Ans.*

(6.) A makes 37 strokes per minute for 5 minutes, and 32 per minute for 5 minutes; B makes 40 strokes per minute for 5 minutes, and 35 per minute for 17 minutes; C, 30 strokes per minute for 12 minutes;

$$\begin{array}{rcrcr} 37 \times 5 = 185 & & 40 \times 5 = 200 & & 30 \\ 32 \times 5 = \underline{160} & & 35 \times 17 = \underline{595} & & \underline{12} \\ 345 & + & 795 & + & 360 = 1500; \end{array}$$

A receives $\frac{345}{1500}$ of $2 = 46 ct., *Ans.*; B, $\frac{795}{1500}$ of $2 = $1.06, *Ans.*; C, $\frac{360}{1500}$ of $2 = 48 ct., *Ans.*

(7.) B has $5600 × 12 = $67200 for 1 mo.; $4200 × 12 = $50400; A must contribute $67200 — $50400 = $16800 for 1 mo., or $16800 ÷ 8 = $2100, for 8 mo.

$2100, *Ans.*

(8.) $4500 × 2 = $9000 capital; $1500 × 3 = $4500; $500 × 3 = $1500; $9000 — $4500 — $1500 — $2200 = $800 loss; A has $4500 for 3 mo., $3000 for 3 mo., and $1500 for 3 mo., in all, $27000 for 1 mo.; B has $4500 for 3 mo., $4000 for 3 mo., $3500 for 3 mo., and $3000 for 3 mo., in all, $45000 for 1 mo.; stock-equivalents are 27000 and 45000, or 3 and 5; hence A should suffer $\frac{3}{8}$, and B $\frac{5}{8}$ of the loss; $\frac{3}{8}$ of $800 = $300; as A should lose $300, and has no capital, he, therefore, owes B $300, that is, B takes the $2200, and has a claim on A for $300, *Ans.*

(9.) A has 12 shares 3 mo., 5, 6 mo., and 9, 3 mo., in all, 93 shares 1 mo.; B has 8 shares 3 mo., 10, 2 mo., and 7, 7 mo., total, 93 shares 1 mo.; C has 7 shares 3 mo., 8, 2 mo., 9, 4 mo., and 7, 3 mo., in all, 94 shares 1 mo.; D has 3 shares 3 mo., 7, 2 mo., 9, 4 mo., and 7, 3 mo., total, 80 shares 1 mo.; $93 + 93 + 94 + 80 = 360$; $\frac{93}{360}$ of $18000 = $4650 each for A and B; $\frac{94}{360}$ of $1800 = $4700 for C; $\frac{80}{360}$ of $18000 = $4000 for D, *Ans.*

(10.) A had $400 for 3 mo., $200 for 9 mo., in all, $3000 for 1 mo.; B, $500 for 4 mo., $200 for 4 mo., $50 for 4 mo., in all, $3000 for 1 mo.; C, $300 for 6 mo., $200 for 6 mo., in all, $3000 for 1 mo.; hence their shares of the profit must be equal. Since A's gain was $225, the entire gain was 3 times $225 = $675, *Ans.*

BANKRUPTCY.

Art. 359.

(2.)

$1200 + $720 + $600 + $1080 = $3600; of this, $2520 is $\frac{7}{10}$; hence, pay 70 ct. on $1, *Ans.*

$\frac{7}{10}$ of $1200 = $840, A; $\frac{7}{10}$ of $720 = $504, B; $\frac{7}{10}$ of $600 = $420, C; $\frac{7}{10}$ of $1080 = $756, D.

(3.)

$16000 less 5% = $15200; of $47500 this sum is $\frac{152}{475} = \frac{8}{25}$; hence, pay $\frac{8}{25}$ of $1, or 32 ct. on $1, and to A. $\frac{8}{25}$ of $3650, or $1168, *Ans.*

ALLIGATION.

Art. 361.

(1.)

Price.		Quality.		Cost.
.80	×	6	=	4.80
.50	×	5	=	7.50
.60	×	5	=	3.00
.40	×	9	=	3.60
		35)		18.90(54 ct., *Ans.*

(2.)

Price.		Quality.		Cost.
$8	×	40	=	$320
10	×	30	=	300
12.50	×	16	=	200
11.75	×	54	=	634.50
		140)		1454.50(10.39, *Ans.*

(3.)

Fineness.		wt.		prod.
16	×	5	=	80
18	×	2	=	36
20	×	6	=	120
24	×	1	=	24
		14)		260($18\frac{4}{7}$ car, *Ans.*

(4.)

wt.		sp. gr.		wt. of water of same bulk.
15	÷	7.75	=	1.93548
8	÷	$6\frac{7}{8}$	=	1.16363
$\frac{1}{4}$	÷	$10\frac{1}{2}$	=	.023809
23.25	÷			3.12292
				= 7.445—, *Ans.*

ANALYSIS.—Copper weighs $7\frac{3}{4}$ times its bulk of water; hence, the bulk of 15 lb. copper is the same as the bulk of 15 ÷ 7.75, or, 1.93548 lb. of water. In like manner 1.16363 lb. water has the bulk of 8 lb. zinc; and .023809 lb. water has the bulk of $\frac{1}{4}$ lb. silver. In all, 3.12292 lb. water has the bulk of $23\frac{1}{4}$ lb. of the combination; hence, the combination weighs $\frac{23.25}{3.12292}$ times its bulk of water, and its specific gravity = 7.445—.

(5.)

%		gal.		prod.
86	×	9	=	774%
92	×	12	=	1104
95	×	10	=	950
98	×	11	=	1078
		42)		39.06(93%, *Ans.*

(6.)

1	branch	50%
1	"	25
1	"	50
1	"	60
1	"	55
1	"	55
6		)295(49,

which, being less than 50, shows *failure.*

ALLIGATION ALTERNATE.

Art. 363. CASE I.

(1.)

			Balance.			lb.
	25	3	17		2	19
	27	1		4		4
28	30	2			3	3 or,
	32	4		1		1
	45	17	3			3
						Ans.

Balance.			lb.
	2	4	6
17			17
	3		3 or,
		3	3
1			1
			Ans.

Balance.			lb.
17			17
	4	2	6
		1	1
	1		1
3			3
			Ans.

(2.)

			Balance.			lb.
	5	$\frac{7}{4}$		1		1
	$5\frac{1}{2}$	$\frac{5}{4}$	5			5
$6\frac{3}{4}$	6	$\frac{3}{4}$			5	5 or,
	7	$\frac{1}{4}$		7		7
	8	$\frac{5}{4}$	5		3	8
						Ans.

Balance.			lb.
5			5
	1		1
		1	1
	5	3	8
7			7
			Ans.

(3.)

			Balance.			Gal.
	84	3	9		1	10
	86	1		7		7
87	88	1			3	3 or,
	94	7		1		1
	96	9	3			3
						Ans.

Balance.			Gal.
7			7
	9	1	10
		1	1
3			3
	1		1
			Ans.

(4.)

OPERATION.

	Bulks per lb.	Diff.	Bal.
$\frac{25}{421}$	$\frac{2}{21}$	$\frac{3487}{97251}$	723
	$\frac{4}{77}$	$\frac{723}{97251}$	3487

EXPLANATION.—Since silver has sp. gr. $\frac{21}{2}$, *one lb.* silver has $\frac{2}{21}$ the bulk of one lb. water; since gold has sp. gr. $\frac{77}{4}$, *one lb.* gold has $\frac{4}{77}$ the bulk of one lb. water; the combination having sp. gr. $\frac{421}{25}$, *one lb.* of the combination must have $\frac{25}{421}$ the bulk of one lb. water. If we take the *whole* in silver, each pound will have $\frac{3487}{97251}$, bulk *too great;* if the whole be taken gold, each pound will have $\frac{723}{97251}$ bulk *too small;* hence, balance in ratio of 723 lb. silver to 3487 lb. gold, *Ans.*

REMARK.—If the sp. gr. of a substance be 4, one pound of it will have a bulk equal to $\frac{1}{4}$ the bulk of a pound of water; if sp. gr. be $\frac{21}{2}$, a pound of the substance will have the bulk of $\frac{2}{21}$ of a pound of water. By thus *inverting* the numbers expressing the sp. gravities of different things, their bulks may be directly compared, just as we compare the *prices per pound* in other examples. When we compare *prices,* in a common example, we balance the losses against the gains, *calling* the units in the balancing, "*pounds,*" though they may be transferred from a column of prices, named in "*cents.*"

ILLUSTRATION.—Suppose, where the average price is $\frac{2}{9}$ ct. we find a *loss* of $\frac{714}{189}$ *cents on one kind,* and gain on *another kind* of a lb. $\frac{24}{189}$ *cents,* we take 714 of the latter kind of *pounds,* and 24 *pounds* of the former. The specific gravity case would read: We lose on a pound $\frac{714}{189}$ *bulks,* and gain on another pound $\frac{24}{189}$ *bulks*; hence, balance by 714 of latter to 24 of former in *pounds.*

$\frac{2}{9}$	$\frac{2}{21}$	$\frac{24}{189}$	714
	$\frac{4}{1}$	$\frac{714}{189}$	24

(5.)

			Bal.		lb.
$\frac{7}{8}$	$\frac{3}{4}$	$\frac{1}{8}$	$\frac{1}{40}$	or,	1
	$\frac{9}{10}$	$\frac{1}{40}$	$\frac{1}{8}$		5

Ans.

(6.)

			Bal.		Parts.
22	24	2	2	1	3
	20	2		1	1
	18	4	1		1

Ans.

CASE II.

Art. 364.

(1.)

			Bal.	
60	50	10	12	$\times 8 = 96$, *Ans.*
	72	12	10	$\times 8 = 80$.

(2.)

			Bal.			
65	40	25	2			$2\times50=100$
	50	15		2		*Ans.* 2
	60	5			2	2
	75	10	5	3	1	$4+(5\times50)=254$, *Ans.*

(3.)

			Bal.	
56	0	56	5	$\times \frac{29}{8} = \frac{145}{8}$ pt. $= 2$ gal. 1 qt. $\frac{1}{8}$ pt., *Ans.*
	91	35	8	$\times \frac{29}{8} = 29$ pt.

(4.)

			gr.	
	16	2	1.8	$\times \frac{81}{1.8} = 3$
18		—		pwt. 9 gr.
	21.6	3.6	1	$\times \frac{81}{1.8} =$

1 pwt. 21 gr., *Ans.*

(5.)

					pt.		
	0	60	3	3	$\frac{9}{5}$	$\frac{12}{5}$	$= \frac{21}{5}$
60	78	18		10		8	pt. =
	96	36	5		3		$4\frac{1}{5}$ pt.

Ans.

(6.)

parts. % acidity.

$12\frac{1}{2}$ pt. $\times$ 0 = 0

$7\frac{1}{2}$ " $\times$ 100 = 750

20)750($37\frac{1}{2}$% obtained.

15

			pt.	
	0	22.5	31	$\times \frac{7\frac{1}{2}}{9} = 25\frac{5}{6}$ pt.
desired, $22\frac{1}{2}$		—		$12\frac{1}{2}$
	100	77.5	9	$13\frac{1}{3}$ pt. =

1 gal. 2 qt. $1\frac{1}{3}$ pt., *Ans.*

(7.)

OPERATION. (1st.)

	1	$\frac{10}{11}$	3			$\times \frac{352}{27} = \frac{352}{9}$	$\frac{352}{9}$ oz.,
1	$\frac{1}{9}$	$\frac{8}{9}$		3	$\frac{1}{2}$		*Ans.*
	4	3	$\frac{10}{11}$	$\frac{8}{9}$	$\frac{8}{54}$	$\frac{640}{54}$	

$12 - \frac{8}{54} = \frac{640}{54}$; $\frac{640}{54} \div \frac{10}{11} = \frac{352}{27}$.

EXPLANATION.—The combination is to *displace* water equal to *once* its weight, and, hence, 1 is the average. The lead, while in the water, displaces $\frac{1}{11}$ of its own weight; the copper displaces $\frac{1}{9}$ of its own weight; the cork, when wholly in water, displaces 4 times its own weight. Hence, the piece, say 1 oz., of lead *lacks* displacing $\frac{10}{11}$ of an oz.; 1 oz. of copper lacks displacing $\frac{8}{9}$; a cork oz. displaces *too much*, by 3 times its weight; hence, balancing, we take 3 oz. of lead for each $\frac{10}{11}$ oz. of cork, and 3 oz. of copper for each $\frac{8}{9}$ oz. of cork. But the conditions require only $\frac{1}{6}$ of 3 oz. of copper; hence, to balance that requires $\frac{1}{6}$ of $\frac{8}{9}$, or $\frac{8}{54}$ oz. cork. The conditions also require 12 oz. cork *in all;* therefore, $12 - \frac{8}{54}$, or $\frac{320}{27}$ oz. cork are yet required, and as this contains $\frac{10}{11}$, $\frac{352}{27}$ times, there must be, by first balancing, $\frac{352}{27} \times 3$, or, $\frac{352}{9}$ oz. lead, which is 2 lb. $7\frac{1}{9}$ oz., *Ans.*

Explanation.—First, balancing in proportion to bulks, 40 of cork to 3 of lead, 32 of cork to 3 of copper. This makes their actual weights, as $40 \times \frac{1}{4}$ to 11×3, and $32 \times \frac{1}{4}$ to 9×3. Then the $\frac{1}{2}$ oz. copper requiring $\frac{8}{54}$ oz. cork, the remaining $\frac{640}{54}$ oz. cork require 2 lb. $7\frac{1}{9}$ oz. lead, *Ans.*

Operation. (2d.)

			propor. bulks.			propor. weights.		
	$\frac{1}{4}$	$\frac{3}{4}$	40	32		10	8	$\frac{8}{54} + \frac{640}{54}$
1	9	8		3	=		27	$\frac{1}{2}$
	11	10	3			33		$\frac{352}{9}$ oz.,

(8.)

$120 \times 74 = 8880$
$150 \times 68 = 10200$
$130 \times 54 = 7020$
$400\ )\ 26100(\ 65\frac{1}{4}\%$.

Since 60 is 120% of that desired, $60 \div 120$, or 50% is the required average.

	$65\frac{1}{4}$	$15\frac{1}{4}$	$10 \times 40 = 400$
50			
	40	10	$15\frac{1}{4} \times 40 = 610$

610 shares, *Ans.*

(9.)

\$

$400 \times 7.50 = 3000$
$640 \times 7.25 = 4640$
$960 \times 6.75 = 6480$
$2000\ )14120(\ 7.06,$ average now.

			bbl.	
	7.06	.56	100	$\times 20 = 2000$
req. av. 6.50				
	5.50	1	56	$\times 20 = 1120$

bbl., *Ans.*

Art. 365.

Case III.

(1.)

The given lbs. and prices make an average of $5\frac{1}{7}$ ct. Then,

	3	3	1		$\frac{3}{4}$ lb., *Ans.*
6	$5\frac{1}{7}$	$\frac{6}{7}$		7	
	7	1	3	6	$6 + 2\frac{1}{4} = 8\frac{1}{4}$ lb., *Ans.*

The balancing requires 1 of the first for 3 of the third; and 7 of the second for 6 of the third. This gives the required 7; but as there are 16 — 13, or

3 yet required, and as these must be taken in proportion, as *one* of the first to *three* of the third, we take $\frac{3}{4}$ of the first balance column and add it to the second; having $\frac{3}{4} + 7 + 8\frac{1}{4} = 16$.

(2.)

bbl.		pr.		$
300	×	7.50	=	2250
800	×	7.80	=	6240
400	×	7.65	=	3060
1500			)	11550($7.70, av. for 1500 bbl.

As there are to be 2000 bbl., there are yet 500 bbl. required, and their price must be $\$(2000 \times 7.85 - 11550) \div 500 = \$8.30$, the av. Hence,

$$8.30 \left| \begin{matrix} 8 \\ 8.50 \end{matrix} \right| \begin{matrix} .30 \\ \hline .20 \end{matrix} \left| \begin{matrix} 2 \\ 3 \end{matrix} \right| \times 100 = \left\{ \begin{matrix} 200 \text{ bbl.} \\ 300 \text{ bbl.} \end{matrix} \right\} \textit{Ans.}$$

(3.)

lb.				$
14	×	.30	=	4.20
20	×	.50	=	10.00
6	×	.60	=	3.60
40)				17.80($44\frac{1}{2}$ ct., av.

56 lb. — 40 lb. = 16 lb. desired.

$(56 \times .40 - 17.80) \div 16 = 28\frac{3}{4}$ ct. av. for 16 lb.

$$28\tfrac{3}{4} \left| \begin{matrix} 25 \\ 35 \end{matrix} \right| \begin{matrix} 3\frac{3}{4} \\ 6\frac{1}{4} \end{matrix} \left| \begin{matrix} 5 \\ 3 \end{matrix} \right| \times 2 = \left\{ \begin{matrix} 10 \text{ lb.} \\ 6 \text{ lb.} \end{matrix} \right\} \textit{Ans.}$$

(4.)

If the specific gravity of a body be $\frac{7}{1}$, it loses $\frac{1}{7}$ in water; so, copper loses $\frac{4}{31}$, and silver $\frac{2}{21}$, while the required loss of combination weight is $\frac{5}{43}$; hence,

$$\frac{5}{43} \left| \begin{matrix} \frac{4}{31} \\ \frac{2}{21} \end{matrix} \right| \begin{matrix} \frac{357}{27993} \\ \hline \frac{589}{27993} \end{matrix} \left| \begin{matrix} \text{oz.} \\ 589 \\ 357 \\ \hline 946 \end{matrix} \right| \times \frac{12}{946} = \left\{ \begin{matrix} 7\frac{223}{473} \text{ oz.} \\ 4\frac{250}{473} \text{ oz.} \end{matrix} \right\} \textit{Ans.}$$

(5.)

1st.

						gr.	
18	15	3	6	2	64	2 pwt. 16 gr., *Ans.*
	20	2		3	24	1 pwt., *Ans.*
	24	6	3		24	1 " *Ans.*
			9	5		

$112 \div 14 = 8.$

Taking 3's and 5's to make 112 gr., we proceed thus to find other answers:

2d.									
	2	2	72	68	64	60	56	52	48
		3	6	15	24	33	42	51	60
	1		34	29	24	19	14	9	4
	3	5							

(6.)

4	9	5	3	2	29	30	31	32	33	34	35	36	37	calves,
	2	2		5	68	60	52	44	36	28	20	12	4	hogs,
	1	3	5		3	10	17	24	31	38	45	52	59	lambs.
			8	7										

Take 5ths of 8 and 5ths of 7 to make 100; or *whole* 8's and *whole* 7's to make 500.

(7.)

If a body have a specific gravity of 2, in water it displaces $\frac{1}{2}$ its own weight; if its sp. gr. be $\frac{4}{3}$, it displaces in like manner $\frac{3}{4}$; so the crown, sp. gr. $\frac{117}{8}$, displaced $\frac{8}{117}$,

$\frac{8}{117}$	$\frac{2}{21}$	$\frac{726}{27027}$	74	$\times \frac{17.5}{195} =$	$6\frac{25}{39}$ *silver.*
	$\frac{4}{77}$	$\frac{444}{27027}$	121		$10\frac{67}{78}$ *gold.*
			195		

and thus with the two metals. Hence, the question is,—*If gold displace, in water, $\frac{4}{77}$ of its own weight, and silver $\frac{2}{21}$ of its own weight, how should these be combined so as to displace $\frac{8}{117}$ of their weight?* The above balancing shows their actual weights should combine as 74 to 121; i. e., the gold should weigh $\frac{121}{195}$ of the *combined* weights. The whole weight being $17\frac{1}{2}$ lb., the weight of the gold must be $17\frac{1}{2}$ lb. $\times \frac{121}{195} = 10\frac{67}{78}$ lb., *Ans.*

INVOLUTION.

Art. 370.

(1.) $(5)^2 = 5 \times 5 = 25$, *Ans.*

(2.) $(14)^3 = 14 \times 14 \times 14 = 2744$, *Ans.*

(3.) $(6)^5 = 6 \times 6 \times 6 \times 6 \times 6 = 7776$, *Ans.*

(4.) $(192)^2 = 192 \times 192 = 36864$, *Ans.*

(5.) $1 \times 1 \times 1 \times 1 \times 1 \times 1 \times 1 \times 1 \times 1 \times 1 = 1$, *Ans.*

(6.) $(\frac{3}{5})^4 = \frac{3}{5} \times \frac{3}{5} \times \frac{3}{5} \times \frac{3}{5} = \frac{81}{625}$, *Ans.*

(7.) $(2\frac{1}{4})^3 = \frac{9}{4} \times \frac{9}{4} \times \frac{9}{4} = \frac{729}{64} = 11\frac{25}{64}$, *Ans.*

(8.) $(\frac{7}{8})^5 = \frac{7}{8} \times \frac{7}{8} \times \frac{7}{8} \times \frac{7}{8} \times \frac{7}{8} = \frac{16807}{32768}$, *Ans.*

(9.) $(.02)^3 = .02 \times .02 \times .02 = .000008$, *Ans.*

(10.) $(5)^4 = 5 \times 5 \times 5 \times 5 = 625$; $\therefore (5^4)^2$, or 5^8, $= 625 \times 625 = 390625$, *Ans.*

(11.) $(.046)^3 = .046 \times .046 \times .046 = .000097336$, *Ans.*

(12.) $(\frac{1}{9})^7 = \frac{1}{9} \times \frac{1}{9} \times \frac{1}{9} \times \frac{1}{9} \times \frac{1}{9} \times \frac{1}{9} \times \frac{1}{9} = \frac{1}{4782969}$, *Ans.*

(13.) $(2056)^2 = 2056 \times 2056 = 4227136$, *Ans.*

(14) $(7.62\frac{1}{2})^2 = 7.62\frac{1}{2} \times 7.62\frac{1}{2} = 58.1406\frac{1}{4}$, *Ans.*

Art. 371.

(1.) $19^2 = (10 + 9)^2 = 100 + 2(10 \times 9) + 81 = 361$, *Ans.*

(2.) $29^2 = (20 + 9)^2 = 400 + 2(20 \times 9) + 81 = 841$, *Ans.*

(3.) $4^2 = (1 + 3)^2 = 1 + 2(1 \times 3) + 9 = 16$, *Ans.*

(4.) $40^2 = (30 + 10)^2 = 900 + 2(30 \times 10) + 100 = 1600$, *Ans.*

(5.) $125^2 = (100 + 25)^2 = 10000 + 2(100 \times 25) + 625 = 15625$, *Ans.*

(6.) $59^2 = (50 + 9)^2 = 2500 + 2(50 \times 9) + 81 = 3481$, *Ans.*

Art. 372.

(1.) $19^3 = (10 + 9)^3 = 1000 + 3(100 \times 9) + 3(10 \times 81) + 729 = 6859$, *Ans.*

(2.) $29^3 = (20 + 9)^3 = 8000 + 3(400 \times 9) + 3(20 \times 81) + 729 = 24389$, *Ans.*

(3.) $4^3 = (1 + 3)^3 = 1 + 3(1 \times 3) + 3(1 \times 9) + 27 = 64$, *Ans.*

(4.) $40^3 = (20 + 20)^3 = 8000 + 3(400 \times 20) + 3(20 \times 400) + 8000 = 64000$, *Ans.*

(5.) $125^3 = (120 + 5)^3 = 1728000 + 3(14400 \times 5) + 3(120 \times 25) + 125 = 1953125$, *Ans.*

(6.) $216^3 = (200 + 16)^3 = 8000000 + 3(40000 \times 16) + 3(200 \times 256) + 4096 = 10077696$, *Ans.*

EVOLUTION.

EXTRACTION OF THE SQUARE ROOT.

Art. 375.

(1.)

$2\dot{8}0\dot{9}$(53, *Ans.*
25
103 | 309
| 309

(2.)

$1\dot{4}4\dot{4}$(38, *Ans.*
9
68 | 544
| 544

(3.)

$\dot{1}1\dot{8}8\dot{1}$(109, *Ans.*
1
209)1881
1881

(4.)

$\dot{1}8\dot{5}6\dot{4}0\dot{6}2\dot{5}$(13625, *Ans.*

```
        1
   23)85
      69
  266)1664
      1596
 2722) 6806
       5444
27245)136225
      136225
```

(5.)

$8\dot{0}0\dot{1}2\dot{3}0\dot{4}$(8944.9—, *Ans.*

```
        64
   169 |1601
       |1521
  1784 |8023
       |7136
 17884 |88704
       |71536
178889 |1716800
       |1610001
          6799
```

(6.)

$\dot{6}2\dot{0}3\dot{7}9\dot{4}$(2490.74, *Ans.*

```
        4
    44 |220
       |176
   489 |4437
       |4401
 49807 |369400
       |348649
498144 |20751
       |19926 —
```

(7.)

$\dot{3}4\dot{4}4\dot{7}3\dot{6}$(1856, *Ans.*

```
       1
  28 |244
     |224
 365 |2047
     |1825
3706 |22236
     |22236
```

(8.)

$\dot{5}7\dot{6}0\dot{0}$(240, *Ans.*

```
      4
  44 |176
     |176
       00
```

(9.)

$1\dot{6}4\dot{9}9\dot{8}4\dot{4}$(4062, *Ans.*

```
      1600
  806 |4998
      |4836
 8122 |16244
      |16244
```

(10.)

$\dot{4}90\dot{9}8\dot{0}4\dot{9}$(7007, *Ans.*
49
14007 | 98049
| 98049

(11.)

$\dot{7}3\dot{0}0\dot{5}$(270.194, *Ans.*
4
47 | 330
| 329
5401 | 10500
| 5401
54029 | 509900
| 486261
540384 | 2363900
| 2161536

(12.)

$386^3 = 5\dot{7}5\dot{1}2\dot{4}5\dot{6}$(7583.69, *Ans.*
49
45 | 851
| 725
1508 | 12624
| 12064
15163 | 56056
| 45489
| 10567
| 9098
1469
1365

(13.)

$\dot{3}.0\dot{0}0\dot{0}0\dot{0}$(1.7320508, *Ans.*
1
27 | 200
| 189
343 | 1100
| 1029
3462 | 7100
| 6924
3464+ | 17600
| 17320+
279
277

(14.)

$\dot{9}.8\dot{6}9\dot{6}0\dot{4}4\dot{0}1\dot{0}$(3.1415926, *Ans.*
9
61 | 86
| 61
624 | 2596
| 2496
6281 | 10004
| 6281
62825 | 372340
| 314125
58215
56543
1672
1257
415
377

(15.) $\sqrt{.030625} = .175$; $\sqrt{40.96} = 6.4$; $\sqrt{.00000625} = .0025$; $.175 \times .0025 \times 6.4 = .0028$, *Ans.*

(16.) $126 \times 58 \times 604 = 4414032$, *Ans.*

(17.) $\sqrt{12.96} \times \sqrt{\frac{5}{6}} = \sqrt{10.8} = 3.2863$, *Ans.*

Art. 377.

(1.) $\sqrt{\frac{6}{7}} = \sqrt{\frac{42}{49}} = \frac{1}{7}\sqrt{42} = .92582+$, *Ans.*

(2.) $\sqrt{34\frac{5}{8}} = \sqrt{34.625} = 5.8843+$, *Ans.*

(3.) $\sqrt{\frac{4}{7}} = \sqrt{\frac{28}{49}} = \frac{1}{7}\sqrt{28}$ and *not* $\frac{1}{7}\sqrt{36}$; hence, $\frac{5}{7}$ more nearly than $\frac{6}{7}$, *Ans.*

(4.) $\sqrt{272.25} = 16.5$, *Ans.*

(5.) $\sqrt{6.40} = 2.5298+$, *Ans.*

(6.) $\frac{28}{57} \times \frac{392}{2527} \times \frac{35}{38} \times \frac{3}{1} = \frac{784}{361 \times 361} \times 35$; hence, sq. rt. $= \frac{28}{361}\sqrt{35} = 5.9160798 \times \frac{28}{361} = .45886+$, *Ans.*

(7.) $\sqrt{123.454321 \times .81} = 11.111 \times .9 = 9.9999$, *Ans.*

(8.) $\sqrt{1.728 \times 4.8 \times \frac{3}{7}} = \sqrt{1.44 \times 1.44 \times 4 \times \frac{3}{7}} = 1.2 \times 1.2 \times 2 \times \frac{1}{7}\sqrt{21} = \frac{2.88}{7}\sqrt{21}$, *Ans.*

EXTRACTION OF THE CUBE ROOT.

Art. 380.

(1.)

$51\dot{2}(8$, *Ans.*
512

(2.) $1\dot{9}68\dot{3}(27$, *Ans.*

```
                        8
  4 × 300 = 1200 | 11683
2 × 7 × 30 =  420 |
    7 ×  7 =   49 |
             1669 |
                  | 11683
```

(3.) $\dot{7}30\dot{1}38\dot{4}$(194, *Ans.*

	1
1×300=300	6301
1×9×30=270	
9×9= 81	
651	5859
361×300=108300	442384
19×4×30= 2280	
4×4= 16	
110596	442384

(4.) $9\dot{4}81\dot{8}81\dot{6}$(456, *Ans.*

	64
16×300=4800	30818
4×5×30= 600	
5×5= 25	
5425	27125
2025×300=607500	3693816
45×6×30= 8100	
6×6= 36	
615636	3693816

(5.)

$\dot{1}06\dot{7}46\dot{2}64\dot{8}$(1022, *Ans.*

	1
1×300=300)	67
100×300=30000	67462
10×2×30= 600	
2× 2= 4	
30604	61208
10404×300=3121200	6254648
102×2×30= 6120	
2× 2= 4	6254648
3127324	

(6.)

$\dot{5}.08\dot{8}44\dot{8}$(1.72, *Ans.*

	1
300	4088
210	
49	3913
559	175448
86700	
1020	
4	
87724	175448

(7.)

$2\dot{2}18\dot{8}.04\dot{1}$(28.1, *Ans.*

	8
1200	14188
480	
64	13952
1744	236041
235200	
840	
1	
236041	236041

(8.)

```
            32.650(3.196154+,
            27           Ans.
     2700| 5650
       90|
        1| 2791
   2791  |2859000
  288300|
    8370|
      81|2670759
  296751  |188241000
 30528300|
    57420|
       36|183514536
 30585756  |4726464
            305858
            166788
            15292
             1386
```

(9.)

```
       .007900(.1991632+,
          1            Ans.
   300 |6900
   270 |
    81 |5859
   651  |1041000
 108300 |
   5130 |
     81 |1021599
 113511    |19401000
  11880300|
      5970|
         1|
  11886271|11886271
             7514729
             713176
              38296
              3565
               264
```

(10)

```
        3.009200(1.443724,
        1            Ans.
   300|2009
   120|
    16|1744
   436| 265200
  58800|
   1680|
     16|241984
  60496| 23216000
 6220800|
   12960|
       9|18701307
 6233769| 4514693
          436364, etc., as above.
```

(11.)

```
 23/729=.031550068587(.315985,
          27                Ans.
     2700| 4550
       90|
        1| 2791
     2791| 1759068
   288300|
     4650|
       25|1464875
   292975  |294193587
  29767500|
     85050|
        81|268673679
  29852631  2551990 8
            2388210, etc., etc.
```

(12.)

```
    2̇5(2.924018, Ans.
     8
1200|17000
 540|
  81|16389
1821|  611000
252300|
  1740|
     4|508088
254044|102912000
25579200|
   35040|
      16|102457024
25614256      454976
              2561
              1988
```

(13.)

```
    1̇1( 2.22398, Ans.
     8
1200 |3000
 120 |
   4 |2648
1324 | 352000
145200 |
  1320 |
     4 |293048
146524 | 58952000
14785200|
   19980|
       9|44415567
14805189|14536433
         1332467
          121176
          11844
```

(14.)

```
2/3=.66̇6(.87358, Ans.
     512
19200 |154666
 1680 |
   49 |146503
20929   |8163666
2270700 |
   7830 |
      9 |6835617
2278539 |1328049
         113927
          18877
          1823
```

(15.)

```
4/15=.26̇6(.64366, Ans.
     216
10800| 50666
  720|
   16| 46144
11536|  4522667
1228800|
   5760|
      9|3703707
1234569| 818960
         74074
          7822
```

(16.) $17\dot{1}.41\dot{6}32\dot{8}87\dot{5}$ (5.555, *Ans.*

125

$25 \times 300 = 7500$ | 46416
$5^2 \times 30 = 750$
$5^2 = 25$
8275 | 41375

5041328
$55^2 \times 300 = 907500$
$55 \times 5 \times 30 = 8250$
$5 \times 5 = 25$
915775 | 4578875

462453875
$555^2 \times 300 = 92407500$
$555 \times 5 \times 30 = 83250$
$5 \times 5 \times 25$
92490775 | 462453875

(17.) $\dot{7}01\dot{1}$ (19.1393267, *Ans.*

1

300
270
81
651

6011
5859

152000
$19^2 \times 300 = 108300$
$19 \times 30 = 570$
1
108871 | 108871

43129000
$191^2 \times 300 = 10944300$
$191 \times 3 \times 30 = 17190$
9
10961499 | 32884497

10244503000
$1913^2 \times 300 = 1097870700$
$1913 \times 30 \times 9 = 516510$
$9 \times 9 = 81$
1098387291 | 9885485619, etc., etc.

(18.) $\sqrt[3]{\frac{48}{4394}} = \sqrt[3]{\frac{24}{2197}} = \frac{1}{13}\sqrt[3]{24} = \frac{1}{13}$ of $2.8844991 =$ $.2218845$, *Ans.*

(19.) $\sqrt[3]{\frac{2}{3} \text{ of } \frac{4}{11}} = \sqrt[3]{.24242424+} = .6235319$, *Ans.*

EXTRACTION OF ANY ROOT.

Art. 384.

(1.)

0	15625(125, *Ans.*
1	1
1	*5625
1	44
*2	*1225
2	1225
22	
2	
*24	
5	
245	

(2.)

0	0	6̇8719̇476̇736̇(4096, *Ans.*
4	16	64
4	16	* 4719476
4	32	4417929
8	*48	* 301547736
4	10881	301547736
*12	490881	
09	10962	
1209	*501843	
9	73656	
1218	*50257956	
9		
*1227		
6		
12276		

(3.)

0	0	0	0	14348907(27, *Ans.*
2	4	8	16	*11148907
4	12	32	*80	11148907
6	24	*80	1592701	
8	*40	113243		
*10	4749			
107				

(4.)

0	0	151(5.325074, *Ans.*
5	25	125
5	50	* 26.000
10	*75	23.877
5	4.59	* 2.123000
*15	79.59	1.691768
.3	4.68	* .431232000
15.3	*84.27	424935125
.3	.3184	* .006296875
15.6	84.5884	595467
.3	.3188	34220
*15.9	*84.9072	
.02	.079825	
15.92	84.987025	
.02	.079850	
15.94	*85.066875	
.02		
*15.96		
.005		
15.965		
15.970		
*15.975		

(5.) Proceeding by Art. 382, we have $\sqrt[4]{97.41}$ = sq. rt.

of $\sqrt{97.41}$, which $= \sqrt{9.86965045} = 3.14159$, or, 3.1416, *Ans.*

(6.) $\sqrt[4]{1.08} = \sqrt{\text{sq. rt. of } 1.08} = \sqrt{1.03923048} = 1.01943$, *Ans.*

(7.)

$\frac{5}{12} = .41\dot{6}$;

0	0	0	0	.4166(.83938, *Ans.*
8	64	512	4096	889866667
16	192	2048	204800000	227626024
24	384	5120000	220746881	9381865
32	64000	5315627	2372916005	1944016
400	65209	5514908	24249351	
403	66427	5717870	24775122	
406	67654	57799	2479283	
409	68890	58419	2481054	
412		59039		
415				

(8.)

$35.2 = 32 \times 1.10$. Hence root $= 2\sqrt[5]{1.10}$.

0	0	0	0	1.10	(1.01924
1	1	1	1	*.10	
2	3	4	*5	*.0489899499	
3	6	*10	5.10100501	*.0013207560	
4	*10.	10.100501	*5.20302005	109721	
*5.01	10.0501	10.201504	5.2965771	22354	
5.02	10.1003	*10.303010	5.390967+	21944	
5.03	*10.1506	10.395228	951	410	
5.04	10.2010	10.487854	5.4860		
*5.05	10.2464	92	&c.,		
&c.,	10.2918	10.579			

$1.01924 \times 2 = 2.03848$, *Ans.*

(9.) $\sqrt[12]{782757789696} = \sqrt[4]{\text{cube rt.}} = \sqrt[4]{9216} = \sqrt{96} =$ 9.79795897, *Ans.*

(10.) $\sqrt[9]{1367631} = \sqrt[3]{\text{cube rt.}} = \sqrt[3]{111} = 4.8058955$, *Ans.*

APPLICATIONS OF SQUARE AND CUBE ROOT.

Art. 388.

(1.) $\sqrt{30^2 + 12^2} = \sqrt{1044} = 32.31+$ ft., *Ans.*

(2.) $\sqrt{10^2 \times 2} = 10\sqrt{2} = 14.142+$ ft., *Ans.*

(3.) $\sqrt{69^2 + 92^2} = \sqrt{13225} = 115$; and $(69 + 92) - 115 = 46$ rd., *Ans.*

(4.) $\sqrt{500^2 + 360^2} = \sqrt{379600} = 616+$ yd., *Ans.*

GENERAL EXERCISES IN EVOLUTION.

Art. 389.

(1.) One side : other :: $\sqrt{12\frac{1}{4}}$: 1; *i. e.*, it is $3\frac{1}{2}$ times the other, *Ans.*

(2.) Larger : smaller :: $12^2 : 5^2$; and $144 \div 25 = 5\frac{19}{25}$, *Ans.*

(3.) $4^3 : 50^3$ :: 16 cu. in. : $125000 \times 16 \div 64 = 31250$ cu. in., *Ans.*

(4.)

By Art. 389, 4,

875 cu. in : 189 cu. in. :: 17.5^3 : cu. of req. no.

Fourth term $= 1157.625$, and $\sqrt[3]{1157.625} = 10\frac{1}{2}$ in., *Ans.*

NOTE.—Evolution having been presented, such examples afford the teacher an opportunity of showing that the same powers, or same roots, of four proportionals are in proportion. This would furnish a very elegant solution:

875 : 189 :: 17.5^3 : cube of no.
125 : 27 :: 17.5^3 : " "
5 : 3 :: 17.5 : the no., $= 10\frac{1}{2}$.

(5.) As it was a perfect power, and the right hand period was 25, the last figure of the root must have been 5; hence, the last trial divisor was $4725 \div 5 = 945$. Then, by the rule, we know 5 to have been annexed to 94, which must have been "double the root already found;" that portion of the root, therefore, must have been $\frac{1}{2}$ of 94, or 47; the entire root, 475, and the power, $475^2 = 225625$, *Ans.* Briefly thus:

$$4725 \div 5 = 945;\ \tfrac{1}{2} \text{ of } 94 = 47;\ 475^2 = 225625, \textit{ Ans.}$$

(6.) $504 = \frac{2}{7}$ of its square; the square $= \frac{7}{2}$ of $504 = 1764$; $\sqrt{1764} = 42$, *Ans.*

(7.) $91252.5 = 1 \times 2.5 \times 3$ times the cube of the smallest. Hence, smallest $= \sqrt[3]{91252.5 \div 7.5} = 23$; hence, 23, $57\frac{1}{2}$, 69, *Ans.*

(8.) *First.*—If the number be an integer it will be the greatest square in 1332; if we extract the root of 1332, to two places, we shall find the integral remainder equal to the integral part of the root. Hence, 36, *Ans.*

Second.—Whether the number sought be integer or fraction, *its square increased by once the number* makes the given sum. Let the diagram for square root (Ray's New Higher) be taken for illustration. The large square A, in the cut, is the square of the number we seek. The rectangle B equals the number multiplied by $\frac{1}{2}$; C equals the same; and *both* equal *once* the number. Now, as in the cut, the width of the rectangle being $\frac{1}{2}$, we want yet a small square, D, the *square* of $\frac{1}{2}$, $= \frac{1}{4}$, to make a complete square $= 1332.25$, the exact root of which is 36.5; but the number we sought is the side of A, which is less by .5 than the root just found; hence 36, *Ans.*

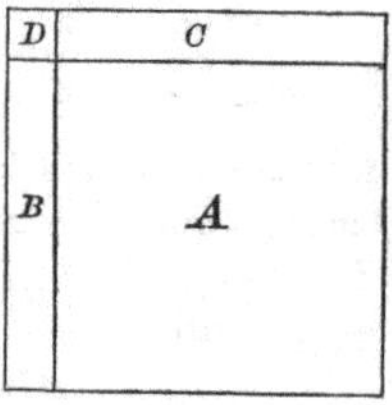

Fig. 1.

(9.) The ceiling is a rectangle of a length equal to $\frac{6}{5}$ of its width, and, if $\frac{1}{6}$ of it be taken off by a line parallel to the end, a square would be left, whose side is the present width. Hence, the ceiling in the given form is $\frac{6}{5}$ of such a square. Let us call this square, E. The *increase* stated is equal to a strip 1 ft. wide, on the *end* of the ceiling, another strip $\frac{6}{5}$ of this, and 1 sq. ft. at the corner. Without this square foot the area will consist of:

$$\left.\begin{array}{l}\frac{6}{5}\text{ of a square E,}\\ \frac{6}{5}\text{ of a foot-wide strip,}\\ \frac{5}{5}\text{ of a foot-wide strip,}\end{array}\right\} = 303 \text{ sq. ft.};$$

or, in all, $\frac{6}{5}$ of a sq. E, $+\ \frac{11}{5}$ of a 1 ft. strip of same length. If we take $\frac{5}{6}$ of this quantity, we shall have *one* sq. E $+\ \frac{11}{6}$ of such a strip.

But $\frac{11}{6}$ of such a strip, *a foot* wide, equals a strip of the same length, $\frac{11}{6}$ of a foot wide; or, it equals *two* strips of that length, $\frac{11}{12}$ of a ft. wide; *i. e.*, 11 inches wide. These two strips can be placed on adjacent sides of the square E; and, a small square of 121 sq. in., at the corner, will then complete a new square having $\frac{5}{6}$ of 303 sq. ft., + 121 sq. in., = 36481 sq. in.; the side of it = $\sqrt{36481} = 191$ in., which is 11 in. more than the required width; hence, 180 in., or 15 ft. = width, and 18 ft. = length, *Ans.*

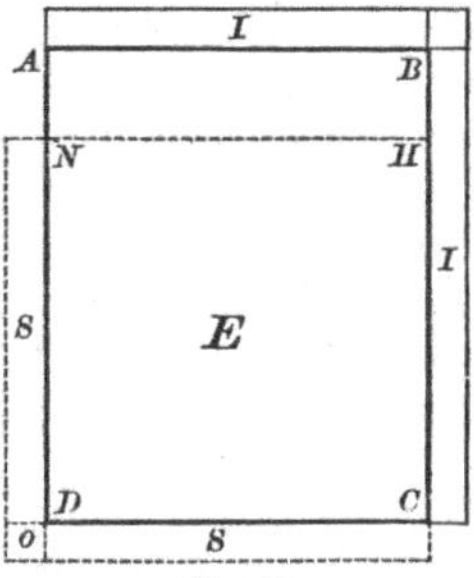

Fig, 2.
A, B, C, D, the ceiling.
I, I, the increase.
N, H, C, D, the square.
S, S, the two 11-inch strips.
O, the 121 sq. in.

(10.) The last three figures, 984, make the last period of the complete power; and the last figure of the root must be 4, for no other of the nine digits will make 4 at the units of the cube. Hence, 241984 ÷ 4, or 60496, must be the complete divisor. This divisor was made by two increases of the trial divisor; the last increase having been $4^2 = 16$, the former part, 60480, = 300 times the *square* of the *partial root* and 4 times 30 times *that root;* hence, $\frac{1}{300}$ of this, or 201.60, must equal *once* the square of that root + $\frac{2}{5}$ of it. But, according to NOTE 3, under the Rule for Square Root, it takes *more* than *twice* an integer added to its square to make the square of the next higher integer; consequently the greatest square in 201 is the square of the partial root already found before the last figure, 4; as this square is 196, and the sq. root is 14, the cube root sought must have been 144, and the power required $= 144^3 = 2985984$, *Ans.*

REMARK.—In every such case where the conditions are that the root and the power are integral, the square of the last figure being taken away, leaves 300 times the square of an integer, + 30 times the integer $\times$ by the last figure. Now, as the last figure can never be greater than 9, this sum can never be more than 300 times the square of an integer, + 270 times the integer; the 300th part of this can never exceed the square of the integer by more than $\frac{270}{300}$, or $\frac{9}{10}$, of the integer, and hence the square of *that integer* MUST BE the *greatest* square in the quotient of this partial trial divisor by 300. The student can convince himself that the 300th part of the *whole* divisor contains *no* square of a greater integral root than *the* root "already found." Hence the following brief operation:

1. $241984 \div 4 = 60496$, the complete divisor.
2. $60496 \div 300 = 201.65+$.
3. The root of greatest square in 201, is 14.
4. The root found was 144, and 144^3 is the answer.

(11.) If one cube differ from another, 1 inch, in the equal dimensions, the difference in solidity will be, one corner cube, = 1 solid inch, and six blocks of the same length as the smaller cube, three of them being *square*, and each of the other three 1 inch wide. Without the corner cube the 3 square blocks and 3 narrow blocks, in this case, will contain 1656 solid inches; or, 1 square block and 1 narrow block contain 552 solid inches. These being

1 inch thick, if we take *one side surface in each* of the two, we have 552 square inches; that is, one square and an inch-wide strip of the same length contain, together, 552 square inches. This surface will be the *same amount*, if the inch-strip be divided into *two half-inch* strips, and these be placed on adjacent sides of the square; there is yet required a square $= \frac{1}{4}$ sq. inch, at the corner, to make a full square, 552.25 sq. in., whose side is $\frac{1}{2}$ inch greater than the length of the strip; $\sqrt{552.25} = 23\frac{1}{2}$; therefore, the side of the square, or edge of the cube, *before* reduction, is $23\frac{1}{2}$ in. $+ \frac{1}{2}$ in. $= 24$ inches; and 24 in. $\times$ 24 in. $\times$ 24 in. $= 13824$ cu. in., *Ans.*

(12.) Conceive one of the *sides* of the room, or one of the *ends*, to be turned down (as upon hinges), to a level with the floor. The corner *to* which the fly travels, may then be considered the diagonally opposite corner of a rectangle. In the one case the sides will be 46 ft. $+ 12\frac{1}{2}$ ft., or $58\frac{1}{2}$ ft. length, and 22 ft. width, and the diagonal 62+ ft.; in the other, the sides will be 22 ft. $+ 12\frac{1}{2}$ ft., or $34\frac{1}{2}$ ft. width, and 46 ft. length, and the diagonal $= \sqrt{1190.25 + 2116} = 57\frac{1}{2}$ ft., *Ans.*

(13.) I. The answer to the first part will be the same as if we required the number of stakes which can be driven, *one* foot apart, on a square of 10 ft., there being just 10 of the allowable divisions in the given side. If we put down 11 on the base line, and, perpendicularly over these, as vertices of equilateral triangles, 10 others, there will be, continuing thus, 11 in the first row, 10 in the second,

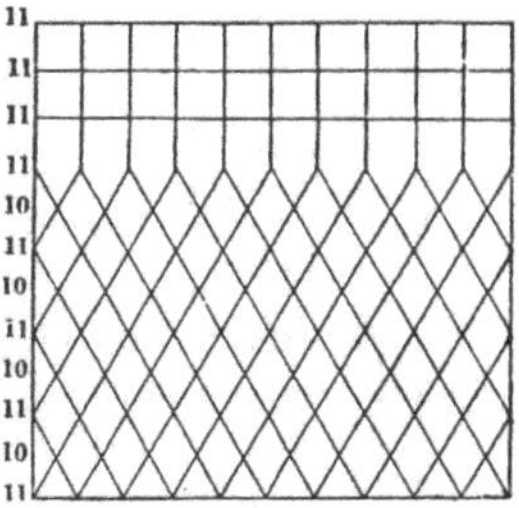

Fig. 3.

11 in the third, 10 in the fourth, and so on. But the *height* of such a triangle is $\sqrt{1-\frac{1}{4}}=.86602$; and in 1 ft. this is not contained more than *once*, so that, on a space 1 foot wide we should have, by triangular arrangement, only $11+10=21$; while by squares we *could* have 22. The triangular arrangement will not be a gain in *two* feet of width unless 2 contain .86602 *oftener* than 2 times; and so, trying, we find that in 6 feet there will be no gain by that method; but in 7 ft. there will be 8 strips, and consequently, 9 rows, by triangular arrangement. In the remaining 3 ft. there will be no gain, and hence we shall observe the square form on that remainder, having

On 7 ft. width	5 rows of 11 = 55	in all, 128 stakes, *Ans.*
	4 " " 10 = 40	
On 3 ft. "	3 " " 11 = 33	

(See Fig. 3.)

REMARK.—If there were 13 rows, of only 10 in each row, the arrangement would be a gain over that above. Let us see if this can be made. The 13 rows would require 12 strips, each $\frac{5}{6}$ ft. wide. If the rows were to reach, from the *base* row, alternately the left hand side, and from the *second* one, alternately the right hand side, there would have to be, for 10 stakes, at least $9\frac{1}{2}$ bases, each equal to $10\div 9.5=\frac{20}{19}$ ft. If so, the half of this, $\frac{10}{19}$ ft., would be a base, and $\frac{5}{6}$ ft., the perpendicular *to allow a hypothenuse of the 1 ft. distance;* but $\sqrt{\frac{100}{361}+\frac{25}{36}}=$ only $\frac{112}{114}+$, which is *not* equal to 1; hence, there can be *no* gain by such arrangement.

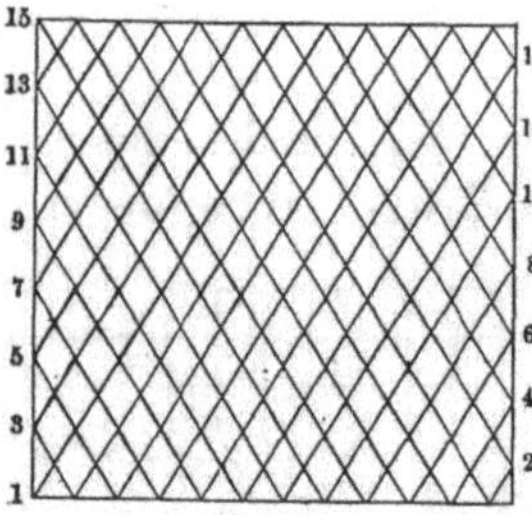

Fig. 4.

II. The answer to the second part is the same as if we required the number of stakes which can be driven, 1 ft.

apart, upon a space 12 ft. square. If we proceed according to the first method shown, we shall have

$$\left.\begin{array}{ll} \text{On 7 ft. width} \left\{\begin{array}{l} 5 \text{ rows of } 13 = 65 \\ 4 \text{ rows of } 12 = 48 \end{array}\right. \\ \text{On 5 ft.} \quad \text{"} \quad\ \ 5 \text{ " } \text{ " } 13 = 65 \end{array}\right\} \text{in all 178 stakes.}$$

But proceeding after the manner indicated in the Remark, let us try to have 15 rows instead of 14. There would be 14 spaces from base to extreme, each $\frac{6}{7}$ ft. wide. Then, making the rows from the base line, alternately reach the left side, and from the second alternately reach the right side, we should need for 12 stakes, $11\frac{1}{2}$ spaces (at least), each $12 \div 11.5$, or $\frac{24}{23}$ ft. Taking half of this for a base, and $\frac{6}{7}$ for a perpendicular, we find the hypothenuse $= \sqrt{\frac{144}{529} + \frac{36}{49}} = \sqrt{\frac{26100}{25921}} = 1 +$, which is more than the allowable distance; hence this arrangement is a gain, and we have 15 rows of 12 stakes, or $15 \times 12 = 180$, *Ans.* (See Fig. 4.)

(14.) For illustration, take any perfect square which is also a perfect cube; *e. g.*, $(2^2)^3 = 64$.

$$\overline{\underline{2 \times 2} \times \underline{2}\ \times}\ \overline{\underline{2} \times \underline{2 \times 2}} = 64.$$

Here it is obvious that the *square* root of the number, is equal to the product of the *cube root by its own square root.* Hence it is easy to arrange the given 5 in the same way, and have $(5^2)^3 = 15625$, *Ans.*

(15.) By attending to the Remark pointed out, the yearly multiplier will be seen to be the product of five equal factors. Hence, the solution is by the extraction of the 5th root of 1.10, which has already been performed under the method of Horner (p. 362, Ex. 8), by separating 35.2 into factors, 1.10×32. See solution in this Key, p. 166; $\sqrt[5]{1.10} = 1.01924$; hence, \$.01924 is the gain on \$1 for the interval; hence, 1.924%, *Ans.*

(16.) (See Art. 386, Remark). The side being a hypothenuse, and the perpendicular being that of a right-angled triangle whose smallest angle is 30°, the perpendicular is such that its *square* is $\frac{3}{4}$ of the square of the side. Hence, the perpendicular $= \frac{1}{2}\sqrt{3}$ times the side. Therefore, the required side $= 4 \div \frac{1}{2}\sqrt{3} = 4 \div .86602$; and the whole boundary, 3 times as much; hence, we have $\$.75 \times (12 \div .86602) = \$10.39$, *Ans.*

SERIES.

ARITHMETICAL PROGRESSION.

Art. 392. CASE I.

(1.) $12 - 1 = 11$; $4 \times 11 = 44$; $44 + 3 = 47$, *Ans.*

(2.) $18 - 1 = 17$; $4 \times 17 = 68$; $100 - 68 = 32$, *Ans.*

(3.) $5\frac{3}{4} - 3\frac{1}{2} = 2\frac{1}{4}$; $2\frac{1}{4} \times 63 = 141\frac{3}{4}$; $141\frac{3}{4} + 3\frac{1}{2} = 145\frac{1}{4}$, *Ans.*

(4.) $.037 - .025 = .012$; $.012 \times 9 = .108$; $.025 + .108 = .133$, *Ans.*

(5.) $71 - 68 = 3$; $3 \times 18 = 54$; $74 - 54 = 20$, *Ans.*

(6.) $130 - 123\frac{1}{2} = 6\frac{1}{2}$; $6\frac{1}{2} \times 5 = 32\frac{1}{2}$; $130 - 32\frac{1}{2} = 97\frac{1}{2}$, *Ans.*

(7.) $12\frac{1}{2} - 6\frac{1}{4} = 6\frac{1}{4}$; $6\frac{1}{4} \times 364 = 2275$; $6\frac{1}{4} + 2275 = 2281\frac{1}{4}$, *Ans.*

Art. 393. CASE II.

(1.) $850 \div 2 = 425$; $425 \times 57 = 24225$, *Ans.*

(2.) $100 + .0001 = 100.0001$; $\frac{100.0001}{2} \times 12345 = 617250.61725$, *Ans.*

(3.) $1 \times 9999 + 1 = 10000$ last term; $10000 + 1 = 10001$; $\frac{10001}{2} \times 10000 = 50005000$, *Ans.*

(4.) $2 \times 999 + 1 = 1999$ last term; $1999 + 1 = 2000$; $\frac{2000}{2} \times 1000 = 1000000$, *Ans.*

(5.) $111 \times 8 = 888$; $999 - 888 = 111$ last term; $999 + 111 = 1110$; $\frac{1110}{2} \times 9 = 4995$, *Ans.*

(6.) $17.25 - 4.12 = 13.13$; $13.13 \times 249 + 4.12 = 3273.49$, last term; $3273.49 + 4.12 = 3277.61$; $\frac{3277.61}{2} \times 250 = 409701.25$, *Ans.*

(7.) $60 - 21 = 39$; $20 - 5 = 15$; $39 \div 15 = 2\frac{3}{5}$ common difference; $2\frac{3}{5} \times 4 = 10\frac{2}{5}$; $21 - 10\frac{2}{5} = 10\frac{3}{5}$, 1st term; $2\frac{3}{5} \times 45 = 117$; $10\frac{3}{5} + 117 = 127\frac{3}{5}$, last term; $10\frac{3}{5} + 127\frac{3}{5} = 138\frac{1}{5}$; $\frac{138\frac{1}{5}}{2} \times 46 = 3178\frac{3}{5}$, *Ans.*

Examples for Practice.

(1.) $(28 - 8) \div (6 - 1) = 4$, *Ans.*

(2.) $(20\frac{3}{4} - 4\frac{1}{2}) \div 13 = 1\frac{1}{4}$, *Ans.*

(3.) $54 - 8 = 46$; $46 \div 2 = 23$; $23 + 8 = 31$, *Ans.*

(4.) $30 - 6 = 24$; $24 \div 6 = 4$; adding 4 to 6, we obtain 10, 14, 18, 22, 26, *Ans.*

(5.) $40 - 4 = 36$; $36 \div 3 = 12$; adding 12 to 4, we obtain 16, 28, *Ans.*

(6.) $3 - 2 = 1$; $1 \div 5 = \frac{1}{5}$. $2\frac{1}{5}$, $2\frac{2}{5}$, $2\frac{3}{5}$, $2\frac{4}{5}$, *Ans.*

(7.) $42 - 9 = 33$; $33 \div 3 = 11$; $11 + 1 = 12$, *Ans.*

(8.) $10\frac{1}{2} - 3 = 7\frac{1}{2}$; $7\frac{1}{2} \div \frac{3}{8} = 20$; $20 + 1 = 21$, *Ans.*

(9.) $500 - 10 = 490$; $\frac{490}{5} = 98$; $98 + 1 = 99$, *Ans.*

(10.) The years of interest on one year's interest on each \$1 of the principal, are equal to the sum of an arithmetical series, 1, 2, 3, etc., to 49, which, by the rule, $= (1 + 49) \times \frac{49}{2} = 1225$ yr. Hence, the amount of each \$1 in the principal, is.

1. The annual interests \$.10 × 50 =	\$5.00
2. The interest of 10 ct. for 1225 yr. =	12.25
3. The dollar itself, ...	1.
In all, ...	\$18.25

\$4927.50 ÷ 18.25 = \$270, *Ans.*

GEOMETRICAL PROGRESSION.

CASE I.

Art. 395.

(1.) $(\frac{1}{2})^{11} = \frac{1}{2048}$; $64 \times \frac{1}{2048} = \frac{1}{32}$, *Ans.*

(2.) $(2\frac{1}{2})^5 = \frac{3125}{32}$; $2 \times \frac{3125}{32} = \frac{3125}{16} = 195\frac{5}{16}$, *Ans.*

(3.) $(\frac{1}{5})^8 = \frac{1}{390625}$; $100 \times \frac{1}{390625} = \frac{4}{15625}$, *Ans.*

(4.) $(3)^9 = 19683$; $4 \times 19683 = 78732$, *Ans.*

(5.) The 3d term, considered the 1st, the 9th would be the 7th; hence, $6^6 = 46656$; $16 \times 46656 = 746496$, *Ans.*

(6.) The 40th term becomes the 8th, starting at the 33d; hence, $(\frac{3}{4})^7 = \frac{2187}{16384}$; $1024 \times \frac{2187}{16384} = 136\frac{11}{16}$, *Ans.*

(7.) Ratio of tran. series $= 2$; $(2)^5 = 32$; $180 \div 32 = 5\frac{5}{8}$, *Ans.*

(8.) Ratio of tran. series $= \frac{2}{3}$; $(\frac{2}{3})^{10} = \frac{1024}{59049}$; $\frac{2187}{4096} \times \frac{1024}{59049} = \frac{1}{4 \times 27} = \frac{1}{108}$, *Ans.*

CASE II.

Art. 396.

(1.) $2^9 = 512$; $6 \times 512 = 3072$, last term; $3072 \times 2 = 6144$; $6144 - 6 \div 1 = 6138$, *Ans.*

(2.) $(\frac{1}{2})^{19} = \frac{1}{524288}$; $16384 \times \frac{1}{524288} = \frac{1}{32}$, last term; $\frac{1}{32} \times \frac{1}{2} = \frac{1}{64}$; $16384 - \frac{1}{64} = 16383\frac{63}{64}$; $16383\frac{63}{64} \div \frac{1}{2} = 32767\frac{31}{32}$, *Ans.*

(3.) $(\frac{2}{3})^6 = \frac{64}{729}$; $\frac{2}{3} \times \frac{64}{729} = \frac{128}{2187}$, last term; $\frac{128}{2187} \times \frac{2}{3} = \frac{256}{6561}$; $\frac{2}{3} - \frac{256}{6561} = \frac{4118}{6561}$; $1 - \frac{2}{3} = \frac{1}{3}$; $\frac{4118}{6561} \div \frac{1}{3} = 1\frac{1931}{2187}$, *Ans.*

(4.) $1 - \frac{1}{2} = \frac{1}{2}$; $1 \div \frac{1}{2} = 2$, *Ans.*

(5.) $1 - \frac{3}{5} = \frac{2}{5}$; $\frac{3}{5} \div \frac{2}{5} = 1\frac{1}{2}$, *Ans.*

(6.) $1 - \frac{3}{4} = \frac{1}{4}$; $\frac{1}{2} \div \frac{1}{4} = 2$, *Ans.*

(7.) $1 - \frac{6}{7} = \frac{1}{7}$; $\frac{7}{6} \div \frac{1}{7} = 8\frac{1}{6}$, *Ans.*

(8.) $1 - \frac{1}{100} = \frac{99}{100}$; $\frac{36}{100} \div \frac{99}{100} = \frac{36}{99} = \frac{4}{11}$, *Ans.*

(9.) $1 - .000001 = .999999$; $.349206 \div .999999 = \frac{22}{63}$; $1 - .001 = .999$; $.480 \div .999 = \frac{160}{333}$; $1 - .1 = .9$; $.6 \div .9 = \frac{2}{3}$.

$\frac{22}{63}$, $\frac{160}{333}$, $\frac{2}{3}$, *Ans.*

(10.) $$\text{Am't} = \frac{\$50\,(156.2472252 - 1)}{1.10 - 1} = \$77623.61, \textit{ Ans.}$$

(11.) The formula of the last example is,

$$\text{Amount} = \frac{\$50\,(1.10^{53} - 1)}{1.10 - 1}.$$

Observe that the second factor in the numerator is the compound amount of \$1, diminished by \$1; *i. e.*, it is the *compound interest* of \$1 for the time. The denominator is .10, the rate; and the quotient of \$50 by the rate is the principal which will yield \$50 each year; *i. e.*, it is the present value of a perpetuity of \$50. This present value, $\times$ comp. int. of \$1 for the time, is exactly what the *Rule* in Case IV requires.

(12.) $\frac{(1.08^1-1)}{.08}=1$. So, using comp. int. table, we have,

$$\frac{1.08^2-1}{.08}=\frac{.1664}{.08}=2.08.$$

$$\frac{1.08^3-1}{.08}=\frac{.259712}{.08}=3.2464.$$

$$\frac{1.08^4-1}{.08}=\frac{.360489}{.08}=4.5061.$$

$$\frac{1.08^5-1}{.08}=\frac{.4693281}{.08}=5.8666.$$

$$\frac{1.08^6-1}{.08}=\frac{.5868743}{.08}=7.3359.$$

Examples for Practice.

(1.) $512 \div 8 = 64$; $\sqrt[3]{64}=4$, *Ans.*

(2.) $49375000000 \div 4\frac{15}{16}=10000000000$; $\sqrt[10]{10000000000}=10$, *Ans.*

(3.) Included terms $=7$; $1000000 \div 729 = \frac{1000000}{729}$; $\sqrt[6]{\frac{1000000}{729}}=\frac{10}{3}=3\frac{1}{3}$, *Ans.*

(4.) No. of terms $=1+2=3$; $112 \div 63 = \frac{112}{63}=\frac{16}{9}$; $\sqrt{\frac{16}{9}}=\frac{4}{3}$; $63 \times \frac{4}{3}=84$, *Ans.*

(5.) No. of terms $=6$; $192 \div 6 = 32$; $\sqrt[5]{32}=2$; multiplying 6 by 2 four times, we have 12, 24, 48, 96, *Ans.*

(6.) No. of terms $=5$; $\frac{1}{9} \div \frac{1}{36864}=4096$; $\sqrt[4]{4096}=8$; multiplying $\frac{1}{36864}$ three times by 8, we have $\frac{1}{4608}$, $\frac{1}{576}$, $\frac{1}{72}$, *Ans.*

(7.) No. of terms $=4$; $3041.28 \div 14.08 = 216$; $\sqrt[3]{216}=6$; multiplying 14.08 twice by 6, we have 84.48 and 506.88, *Ans.*

MENSURATION.

Art. 404.

(1.) 9 ft. 4 in. = 112 in.; 2 ft. 5 in. = 29 in.; 112 in. $\times$ 29 in. = 3248 sq. in. = 22 sq. ft. 80 sq. in., *Ans.*

(2.) 5 ft. 8 in. = $5\frac{2}{3}$ ft.; 42 ft. $\times$ $5\frac{2}{3}$ ft. = 238 sq. ft. = $26\frac{4}{9}$ sq. yd., *Ans.*

(3.) 48 ft. $\times$ 10 ft. = 480 sq. ft. = 69120 sq. in.; 8 in. $\times$ 8 in. = 64 sq. in.; 69120 sq. in. $\div$ 64 sq. in. = 1080, *Ans.*

(4.) 72 rd. $\times$ 16 rd. $\times$ $\frac{1}{2}$ = 576 P. = 3 A. 96 sq. rd., *Ans.*

(5.) 13 ft. 3 in. = 159 in.; 9 ft. 6 in. = 114 in.; 159 in. $\times$ 114 in. $\times$ $\frac{1}{2}$ = 9063 sq. in. = 62 sq. ft. 135 sq. in., *Ans.*

	(6.)				(7.)		
22 in.	42	42	42	15 rd.	29	29	29
24 in.	22	24	38	18 rd.	15	18	25
38 in.	20	18	4	25 rd.	14	11	4
2)84 in.				2)58 rd.			
42				29			

(6.) $42 \times 20 \times 18 \times 4 = 60480$; $\sqrt{60480} = 246 -$ sq. in. = 1 sq. ft. 102 — sq. in., *Ans.*

(7.) $29 \times 14 \times 11 \times 4 = 17864$; $\sqrt{17864} = 133.66$ sq. rd.—, *Ans.*

(8.) 9 ft. + 21 ft. = 30 ft.; $\frac{30}{2}$ ft. $\times$ 16 ft. = 240 sq. ft., *Ans.*

(9.) 43 rd. + 65 rd. = 108 rd.; $\frac{108}{2}$ rd. $\times$ 27 rd. = 1458 sq. rd. = 9 A. 18 sq. rd., *Ans.*

(10.) $10 \times 9 \div 2 = 45$; $12 \times 15 \div 2 = 90$; $16 \times 10\frac{1}{2} \div 2 = 84$; 45 sq. rd. + 90 sq. rd. + 84 sq. rd. = 219 sq. rd. = 1 A. 59 sq. rd., *Ans.*

(11.)

10	20	20	20		14	24	24	24
12	10	12	18		16	14	16	18
18	10	8	2		18	10	8	6
2)40					2)48			

$20 \times 10 \times 8 \times 2 = 3200$; $24 \times 10 \times 8 \times 6 = 11520$; $\sqrt{3200} = 56.568+$ sq. rd. $\sqrt{11520} = 107.331+$, sq. rd. 56.568rd. + 107.331 + rd. = 163.9 — rd. = 1 A. 3.9— sq. rd., *Ans.*

(12.) 25 ft. × $14\frac{1}{4}$ ft. × 2 = $712\frac{1}{2}$ sq. ft. in sides; 18 ft. × $14\frac{1}{4}$ ft. × 2 = 513 sq. ft. in ends; $712\frac{1}{2} + 513 = 1225\frac{1}{2}$ sq. ft., total; $7\frac{1}{6}$ ft. × $3\frac{1}{3}$ ft. = $23\frac{8}{9}$ sq. ft.; $5\frac{2}{3}$ ft. × $3\frac{1}{2}$ ft. × 2 = $39\frac{2}{3}$ sq. ft.; $6\frac{1}{3}$ ft. × $5\frac{1}{2}$ ft. = $34\frac{5}{6}$ sq. ft., $23\frac{8}{9}$ sq. ft. + $39\frac{2}{3}$ sq. ft. = $63\frac{5}{9}$ sq. ft.; one half of $63\frac{5}{9}$ sq. ft. = $31\frac{7}{9}$ sq. ft.; $31\frac{7}{9}$ sq. ft. + $34\frac{5}{6}$ sq. ft. = $66\frac{11}{18}$ sq. ft.; $1225\frac{1}{2}$ sq. ft. — $66\frac{11}{18}$ sq. ft. = $1158\frac{8}{9}$ sq. ft. = $128\frac{62}{81}$ sq. yd., *Ans.*

(13.) $86.6025 \div \frac{10}{2} = 17.3205$, the other diagonal. These being at rt. angles, $\frac{1}{2}$ of 17.3205 is the base, and 5 the perp. of a rt. ang. triangle; hence, $\sqrt{8.66025^2 + 25} \times 4$ = perim., which is $\sqrt{99.9999+} \times 4 = 40$ rd., nearly, *Ans.*

(14.) 48 ft. × 27 ft. × 3 = 3888 sq. ft. = 432 sq. yd.; 12 ft. × $8\frac{1}{4}$ ft. × 3 = 297 sq. ft. = 33 sq. yd.; 432 sq. yd. — 33 sq. yd. = 399 sq. yd. *between* walls; 48 ft. × $\frac{3}{4}$ ft. × 2 × 3 = 216 sq. ft. = 24 sq. yd. *in* the walls; \$1.46 × 399 = \$582.54; 76 ct. × 24 = \$18.24; \$582.54 + \$18.24 = \$600.78, *Ans.*

(15.) The diagonal of a square whose side = 1, is 1.41421; *i. e.*, it exceeds the side by .41421; and, as squares are similar figures,

.41421 : 1 : : given excess : req. side.

.4142 : 1 : : 20.71 ch. : 50 ch.

50 × 50 = 2500; 2500 sq. ch. = 250 A., *Ans.*

(16.)

30 ft. $\times$ 12 ft. $\times$ 2 = 720 sq. ft., sides;
25 ft. $\times$ 12 ft. $\times$ 2 = 600 sq. ft., ends;
30 ft. $\times$ 25 ft. = 750 sq. ft., ceiling;
2070
132
9)1938
$215\frac{1}{3}$ sq. yd., *Ans.*

15 in. $\times$ 2 = $2\frac{1}{2}$ ft.;
5 ft. — $2\frac{1}{2}$ ft. = $2\frac{1}{2}$ ft.;
$8\frac{1}{6}$ ft. $\times$ $2\frac{1}{2}$ ft. $\times$ 3 = $61\frac{1}{4}$ sq. ft.;
7 ft. $\times$ $3\frac{1}{2}$ ft. $\times$ 2 = 49 sq. ft.:
$4\frac{1}{2}$ ft. $\times$ $4\frac{5}{6}$ ft. = $21\frac{3}{4}$ sq. ft;
Deductions, 132 sq. ft.

25 ct. $\times$ $215\frac{1}{3}$ = \$53.83, *Ans.*

(17.) The side of the field is $\sqrt{400} = 20$ ch. The line must cut off either a triangle or a trapezoid. If a *triangle*, it could be no *greater* than a base $20 \times \frac{1}{2}$ a height 8, and this is *less* than $19\frac{1}{2}$ A. Hence, the line must cut off a trapezoid; the height or breadth between parallels = 20 ch.; and $195 \div 20 = 9.75$, the *half sum* of the sides; hence, *sum* of sides = $19\frac{1}{2}$, and $19\frac{1}{2} - 8 = 11\frac{1}{2}$, the side not given. Therefore, the line would be $\sqrt{20^2 + (11\frac{1}{2} - 8)^2}$ if 8 be taken as one of the sides. If the piece cut off be on the side of the square *not* containing the 8, the sides of the trapezoid would be 12, and $19.5 - 12$; in that case the line would be a hypothenuse to base 20, and height $20 - 8 - 7\frac{1}{2}$, or $4\frac{1}{2}$; *i. e.*, hyp. $= \sqrt{400 + 20.25} = 20\frac{1}{2}$ ch., *Ans.*

(18.) 20 ft. $\times$ $10\frac{1}{3}$ ft. $\times$ 2 = $413\frac{1}{3}$ sq. ft.; $14\frac{1}{2}$ ft. $\times$ $10\frac{1}{3}$ ft. $\times$ 2 = $299\frac{2}{3}$ sq. ft.; $4\frac{1}{3}$ ft. $\times$ 4 ft. = $17\frac{1}{3}$ sq. ft.; 6 ft. $\times$ $3\frac{1}{6}$ ft. $\times$ 2 = 38 sq. ft.; $413\frac{1}{3}$ sq. ft. + $299\frac{2}{3}$ sq. ft. = 713 sq. ft.; $17\frac{1}{3}$ sq. ft. + 38 sq. ft. = $55\frac{1}{3}$ sq. ft.; 713 sq. ft. — $55\frac{1}{3}$ sq ft. = $657\frac{2}{3}$ sq. ft. = $73\frac{2}{27}$ sq. yd., *Ans.*

(19.) This makes a surface, $3\frac{5}{6}$ ft. $\times$ 4 = $15\frac{1}{3}$ ft. wide, and 7 ft. 8 in. + 6 ft. 10 in. + 5 ft. 3 in. = $19\frac{3}{4}$ ft. high; $15\frac{1}{3}$ ft. $\times$ $19\frac{3}{4}$ ft. = $302\frac{5}{6}$ sq. ft.; 20 ct. $\times$ $302\frac{5}{6}$ = \$$60.56\frac{2}{3}$, *Ans.*

THE CIRCLE.

Art. 409.

NOTE.—Observe that 3.14159265 is very nearly the same as $\frac{355}{113}$.

(1.) 16 $\times$ 3.14159265 $=$ 50.265482, *Ans.;* $22\frac{1}{4}$ $\times$ 3.14159265 $=$ 69.900436, *Ans.;* 72.16 $\times$ 3.14159265 $=$ 226.6973, *Ans.;* 452 yd. $\times$ $\frac{355}{113}$ $=$ 1420 yd., *Ans.*

(2.) 56 $\div$ 3.14159265 $=$ 17.82539, *Ans.;* $182\frac{1}{2}$ $\div$ 3.14159265 $=$ 58.09, *Ans.;* 316.24 $\div$ 3.14159265 $=$ 100.66232, *Ans.;* 639 ft. $\div$ $\frac{355}{113}$ $=$ 113 $\times$ 1.8 $=$ 203.4 ft.. *Ans.*

(3.) 2 ft. 5 in. $=$ 29 in.; 13 yd. 1 ft. $=$ 40 ft.; 5 ft. $\times$ 5 ft. $\times$ 3.14159265 $=$ 78.54 sq. ft., *Ans;* $14\frac{1}{2}$ in. $\times$ $14\frac{1}{2}$ in. $\times$ 3.14159265 $=$ 660.52 sq. in., *Ans.;* 20 ft. $\times$ 20 ft. $\times$ 3.14159265 $=$ 1256.637 sq. ft. $=$ 139 sq. yd. 5.637 sq. ft., *Ans.*

(4.) 7 ft. 3 in. $=$ 87 in.; 6 yd. 1 ft. 4 in. $=$ $19\frac{1}{3}$ ft.; 23 ft. $\times$ 23 ft. $\div$ 3.14159265 $=$ 168.386 sq. ft., *Ans.;* $43\frac{1}{2}$ in. $\times$ $43\frac{1}{2}$ in. $\div$ 3.14159265 $=$ 602.322 sq. in. $=$ 4 sq. ft. 26.322 sq. in., *Ans.;* $9\frac{2}{3}$ ft. $\times$ $9\frac{2}{3}$ ft. $\div$ 3.14159265 $=$ 29.7443 sq. ft., *Ans.*

(5.) 47.124 $\times$ $3\frac{3}{4}$ $=$ 176.715 sq. ft., *Ans.*

(6.) The fraction is that part of the log which a segment of a circle, height $\frac{1}{3}$ of diameter, is of the whole circle. Let the diameter be 3; the radius, then, $=$ 1.5, the distance from center to base of segment $=$ .5; the half chord $=\sqrt{1.5^2-.5^2}=\sqrt{2}=1.4142136$. The whole base of segment (or whole chord) $=2\sqrt{2}$. The area of the segment, by No. 2, under V, p. 398, is,

$$\frac{1}{4\sqrt{2}}+\tfrac{2}{3}\text{ of }2\sqrt{2}\times 1=2.06239$$

Area of circle $=$ 9 $\times$ .7854, and 2.06239 $\div$ (9 $\times$ .7854) $=$ .2918, *Ans.*

(7.) The fraction required is the part which a *square*

is of the *circle in which it is inscribed.* The diameter being 1, the side of the square $= \frac{1}{2}\sqrt{2}$; the *area* of the circle $= .785398$; the area of the square $= .5$; the fraction sought $= .5 \div .785398 = .6366$, *Ans.*

MENSURATION OF SOLIDS.

Art. 410.

(1.) $5\frac{1}{4} + 6\frac{1}{2} + 8\frac{3}{4} + 10\frac{1}{2} + 9 = 40$; $11\frac{1}{4}$ in. $\times$ 40 in. $=$ 450 sq. in., *Ans.*

(2.) $1\frac{3}{4}$ ft. $=$ 21 in.; 1 ft. $2\frac{1}{2}$ in. $= 14\frac{1}{2}$ in.; $14\frac{1}{2}$ in, $\times$ 3.14159265 $=$ 45.5531 in.; 45.5531 in. $\times$ 21 $=$ 956.6 sq. in. $=$ 6 sq. ft. 92.6 sq. in., *Ans.*

(3.)

$$\begin{array}{r} 60 \\ 80 \\ 100 \\ \hline 2)\overline{240} \\ \hline 120 \end{array} \qquad \begin{array}{c} 120 \\ \underline{60} \\ 60 \end{array} \times \begin{array}{c} 120 \\ \underline{80} \\ 40 \end{array} \times \begin{array}{c} 120 \\ \underline{100} \\ 20 \end{array} \times 120 = 5760000;$$

$\sqrt{5760000} = 2400$ sq. ft.; $2400 \times 2 = 4800$ sq. ft.;

60 + 80 + 100 $=$ 240 ft.; 240 ft. $\times$ 90 ft. + 4800 sq. ft. $=$ 26400 sq. ft., *Ans.*

(4.) 28 ft. $\times$ 19 ft. $=$ 532 sq. ft; $(\frac{19}{2})^2 \div 3.14159265 =$ 28.7275 sq. ft. in one base; 28.7275 sq. ft. $\times$ 2 $=$ 57.455 sq. ft.; 532 + 57.455 $=$ 589.455 sq. ft., *Ans.*

(5.) $640 \times 4 \times \frac{291}{2} = 500480$ sq. ft. conv. surf., *Ans.;* $(640)^2 = 409600$ sq. ft. base; 500480 + 409600 $=$ 910080 sq. ft., *Ans.*

(6.) 66 ft. 8 in. $=$ 800 in.; 4 ft. 2 in. $=$ 50 in.; 50 in. $\times$ 2 $\times$ 3.14159265 $=$ 314.159265 in. circ.;
800 in. $\times$ 314.159265 in. $\div$ 2 $=$ 125663.706 sq. in., convex surface, *Ans.;* $(50)^2 \times 3.14159265 = 7853.9816$ sq. in. in base; 125663.706 + 7853.9816 $=$ 133517.6876 sq. in., *Ans.*

(7.) 1 ft. 2 in. $=$ 14 in.; $4\frac{1}{2}$ in. $\times$ $4\frac{1}{2}$ in. $=$ $20\frac{1}{4}$ sq. in. in base; $20\frac{1}{4}$ sq. in. $\times$ $\frac{14}{3}$ in. $=$ $94\frac{1}{2}$ cu. in., *Ans.*

(8.)
$$\begin{array}{rr} 12 & 18 \\ 12 & \underline{12} \\ \underline{12} & 6 \\ 2)\underline{36} & \\ 18 & \end{array}$$

$18 \times 6 \times 6 \times 6 = 3888$; $\sqrt{3888} = 62.354$ sq. in.; $\frac{1}{3}$ of 15.24 in. $=$ 5.08 in.; 62.354 sq. in. $\times$ 5.08 in. $=$ 316.76 cu. in., *Ans.*

(9.) 9 in. $\times$ 9 in. $=$ 81 sq. in.; 81 sq. in. $\times$ 19 in. $=$ 1539 cu. in., *Ans.*

(10.) $6\frac{1}{2}$ ft. $=$ 78 in.; 2 ft. 10 in. $=$ 34 in.; 1 ft. 8 in. $=$ 20 in.; 34 in. $\times$ 20 in. $\times$ 78 in. $=$ 53040 cu. in. $=$ 30 cu. ft. 1200 cu. in., *Ans.*

(11.) 8 in. $\times$ 12 in. $\div$ 2 $=$ 48 sq. in.; 48 sq. in. $\times$ 7 in. $=$ 336 cu. in., *Ans.*

(12.)
$$\begin{array}{rrrr} 2 & 3\frac{3}{4} & 3\frac{3}{4} & 3\frac{3}{4} \\ 2\frac{1}{2} & \underline{2} & \underline{2\frac{1}{2}} & \underline{3} \\ \underline{3} & 1\frac{3}{4} & 1\frac{1}{4} & \frac{3}{4} \\ 2)\underline{7\frac{1}{2}} & & & \\ 3\frac{3}{4} & & & \end{array}$$

$1\frac{3}{4} \times 1\frac{1}{4} \times \frac{3}{4} \times 3\frac{3}{4} = \frac{1575}{256}$; $\sqrt{\frac{1575}{256}} = \frac{39.69}{16} = 2.48$; $2.48 \times 4\frac{1}{3} = 10.75$ cu. ft. *Ans.*

(13.) $\frac{5}{2} \times \frac{5}{2} \times 3.14159265 = 19.635$ sq. in. in base; 19.635 sq. in. $\times$ $10\frac{1}{2}$ in. $=$ 206.167 cu. in., *Ans.*

(14.) 4 in. $\times$ 4 $=$ 16 in.; $2\frac{1}{2}$ in. $\times$ 4 $=$ 10 in.; (16 $+$ 10) $\div$ 2 $=$ 13 in.; 13 in. $\times$ $3\frac{1}{3}$ in. $=$ $43\frac{1}{3}$ sq. in., *Ans.*

(15.) 7 in. $\times$ 3.14159265 $=$ 21.99114855 in.; 3 in. $\times$ 3.14159265 $=$ 9.42477795 in.; 21.99115 $+$ 9.42478 $\div$ 2 $=$ 15.70796 in.; 15.70796 in. $\times$ 5 in. $=$ 78.5398 sq. in., convex surf., *Ans.*; $(\frac{7}{2})^2 \times 3.14159265 = 38.48451$ sq. in.; $(\frac{3}{2})^2 \times 3.14159265 = 7.06858$ sq. in.; 78.5398 $+$ 38.48451 $+$ 7.06858 $=$ 124.0929 sq. in., whole surface, *Ans.*

(16.) $\frac{1}{3}$ of 1 ft. $4\frac{1}{2}$ in. $=$ $5\frac{1}{2}$ in.; $(10\frac{2}{3})^2 = 113\frac{7}{9}$ sq. in. in lower base; $(4\frac{1}{4})^2 = 18\frac{1}{16}$ sq. in. in upper base; mean base $= \sqrt{(10\frac{2}{3})^2 \times (4\frac{1}{4})^2} = 10\frac{2}{3} \times 4\frac{1}{4} = 45\frac{1}{3}$ sq. in.; $113\frac{7}{9} + 18\frac{1}{16} + 45\frac{1}{3} = 177\frac{25}{144}$ sq. in.; $177\frac{25}{144} \times 5\frac{1}{2} = 974\frac{131}{288}$ cu. in., *Ans.*

(17.) $(9)^2 \times 3.14159265 = 254.469$; $(5)^2 \times 3.14159265 = 78.53982$; mean base $= 9 \times 5 \times 3.14159265 = 141.37167$; $254.469 + 78.53982 + 141.37167 = 474.38049$; $474.38049 \times \frac{16}{3} = 2530.03$ cu. in., *Ans.*

(18.) $(27)^2 \times 3.14159265 = 2290.221+$ sq. ft., *Ans.*; $(10)^2 \times 3.14159265 = 314.16$ sq. in., *Ans.*

(19.) 113.097335 sq. mi. $\times$ 6 mi. $\div 6 = 113.097335$ cu. mi., *Ans.*

(20.) $(4)^3 \times 3.14159265 \div 6 = 33.5103$ cu. ft., *Ans.*

(21.) $40115 \div \frac{355}{113} = 113 \times 113$; $\sqrt{113 \times 113} = 113$ mi. diameter; $40115 \times 113 \div 6 = 755499\frac{1}{6}$ cu. mi., *Ans.*

(22.) The diameter is the diagonal of the cube. If the edge be 1, the diagonal of a face $= \sqrt{2}$, the diagonal of the cube $= \sqrt{2+1} = 1.7320508$, and $1 \div 1.7320508 = .57735$, *Ans.*

MASONS' AND BRICKLAYERS' WORK.

Art. 412.

(1.) $18 \times 8\frac{1}{2} \times 6\frac{1}{6} = 943\frac{1}{2}$ cu. ft.; $943\frac{1}{2} \div 25 = 37.74$, or $37\frac{3}{4}$ perches, nearly, *Ans.*

(2.) $20 \times 7\frac{3}{4} \times 2 = 310$ cu. ft.; this $= 12\frac{52}{99}$ perches; $\$\frac{3}{4} \times 12\frac{52}{99} = \$\frac{930}{99} = \$9.39$, *Ans.*

(3.) 22 ft. 5 in. — twice 1 ft. 10 in. = 18 ft. 9 in., length of end wall; $36 + 36 + 18\frac{3}{4} + 18\frac{3}{4} - 8 = 101\frac{1}{2}$ ft.; $101\frac{1}{2} \times 9\frac{1}{3} \times 1\frac{5}{6} \div 24\frac{3}{4} = 70\frac{14}{81}$ perches; $\$2.75 \times 70\frac{14}{81} = \$192.98$, *Ans.*

(4.) 1 brick is $8 \times 4 \times 2 = 64$ cu. in. $= \frac{1}{27}$ cu. ft.; $150 \times 8\frac{1}{2} \times 1\frac{1}{3} = 1700$ cu. ft.; $1700 \times \frac{9}{10} \div \frac{1}{27} = 41310$ bricks $= 41.31$ thousand; $\$7 \times 41.31 = \$289.17$, *Ans.*

(5.) $(10)^2 = 100$; $(4)^2 = 16$; mean $= \sqrt{(10)^2 \times (4)^2} = 40$; $100 + 16 + 40 = 156$; $156 \times \frac{86}{3} = 4472$ cu. ft., whole chimney; $3 \times 3 \times 86 = 774$ cu. ft. in cavity; $4472 - 774 = 3698$ cu. ft. of brick work; $3698 \times \frac{9}{10} \div \frac{1}{27} = 89861+$ bricks, *Ans.*

GAUGING.

Art. 414.

(1.) $99 \times 41 \times 34 \div 2150.42 = 64.18$ bu., *Ans.*

(2.) $(\frac{13}{2})^2 \times 3.14159265 = 132.7323$; $(\frac{10}{2})^2 \times 3.14159265 = 78.5398$; mean $= \frac{13}{2} \times \frac{10}{2} \times 3.14159265 = 102.1018$; $132.7323 + 78.5398 + 102.1018 = 313.3739$; $313.3739 \times \frac{12}{3} = 1253.4956$ cu. in.; 1253.4956 cu. in. $\div 231 = 5.4264$ gal., *Ans.*

(3.) 11 ft. 6 in. $= 138$ in.; 7 ft. 8 in. $= 92$ in.; $92 \times 92 \times .0034 \times 138 = 3971.3088$ gal.; $3971.3088 \div 31.5 = 126.0733$ bbl., *Ans.*

(4.) 10 ft. $= 120$ in., 9 ft. $= 108$ in., 5 ft. $= 60$ in.; $(120)^2 = 14400$; $(108)^2 = 11664$; mean $= 120 \times 108 = 12960$; $14400 + 11664 + 12960 = 39024$; $39024 \times \frac{60}{3} = 780480$ cu. in.; $780480 \div 231 \div 31.5 = 107.26$ bbl., *Ans.*

Art. 415.

(1.) This is a frustum of a cone; $(\frac{16}{2})^2 \times 3.14159265 = 201.062$; $(\frac{13}{2})^2 \times 3.14159265 = 132.732$; mean base $= \frac{16}{2} \times \frac{13}{2} \times 3.14159265 = 163.363$; $201.062 + 132.732 + 163,363 = 497.157$; $497.157 \times \frac{26}{3} = 4308.694$ cu. in.; 4308.694 cu. in. $\div 231 = 18.65$ gal., *Ans.*

(2.) Difference of the diameters $= 3$ in.; $\frac{6}{10}$ of 3 in. $=$ 1.8 in.; $18 + 1.8 = 19.8$; $19.8 \times 19.8 \times 34 \times .0034 =$ 45.32 gal., *Ans.*

(3.) Difference of the diameters $= 6$ in.; $\frac{2}{3}$ of 6 in. $=$ 4 in.; 3 ft. $+$ 4 in. $= 40$ in.; $40 \times 40 \times 64 \times .0034 =$ 348.16 gal., *Ans.*

LUMBER MEASURE.

Art. 416.

(2.) $(24 - 4)^2 \times \frac{12}{16} = 300$ ft., *Ans.*

(3.) $(25 - 4)^2 \times \frac{24}{16} = 661\frac{1}{2}$ ft., *Ans.*

(4.) $(50 - 4)^2 \times \frac{12}{16} = 1587$ ft., *Ans.*

MEASURING GRAIN AND HAY.

Art. 419.

(1.) $10\frac{1}{2} \times 3\frac{1}{2} \times 2 = \frac{21}{2} \times \frac{7}{2} \times 2 = 73.5$ cu. ft.;

$$\left.\begin{array}{l} 73.5 \times .8 = 58.8 \text{ bu.} \\ 73.5 \times .8 \times \frac{1}{2} = 29.4 \text{ bu.} \\ 73.5 \times .8 \times \frac{1}{3} = 19.6 \text{ bu.} \end{array}\right\} \textit{Ans.}$$

(2.) $40 \times 16 \times 10 \times .8 = 5120$ bu., *Ans.*

(3.) $48000 \div 550 = 87\frac{3}{11}$ tons clover, *Ans.*; and $48000 \div 450 = 106\frac{2}{3}$ tons timothy, *Ans.*

MISCELLANEOUS EXERCISES.

(1.) I would gain 2 ct. more a dozen, which is $\frac{1}{6}$ of a cent apiece; $\frac{1}{4} + \frac{1}{6} = \frac{5}{12}$ ct., *Ans.*

(2.) 3 for a dime is $3\frac{1}{3}$ ct. each; $3\frac{1}{3}$ ct. $-$ $\frac{1}{2}$ ct. $= 2\frac{5}{6}$ ct. cost; 4 for a dime is $2\frac{1}{2}$ ct. apiece; $2\frac{5}{6}$ ct. $-$ $2\frac{1}{2}$ ct. $=$ $\frac{1}{3}$ ct., *Ans.*

(3.) The difference between $37\frac{1}{2}$ ct. and 45 ct., or $7\frac{1}{2}$ ct., is $\frac{2}{5}$ of the gain at 45 ct.; $7\frac{1}{2}$ ct. is $\frac{2}{5}$ of $18\frac{3}{4}$ ct., the gain per bushel at 45 ct.; 45 ct. $-$ $18\frac{3}{4}$ ct. $= 26\frac{1}{4}$ ct. a bu., *Ans.*

(4.) The difference of gain is 15 ct.; there must be as many oranges as $\frac{3}{8}$ ct. is contained times in 15 ct.; $15 \div \frac{3}{8} = 40$ oranges, *Ans.*

(5.) By gaining 11 ct. I receive 27 ct. more than by losing 16 ct.; the difference in price is 3 ct. a doz.; there are $27 \div 3 = 9$ doz., costing 5 ct. $\times 9 + 16$ ct. $= 61$ ct.; 61 ct. $\div 9 = 6\frac{7}{9}$ ct., *Ans.*

(6.) I must charge $\frac{2}{3} + \frac{2}{3} = \frac{4}{3}$ ct. more apiece; this is $\frac{4}{3}$ ct. $\times 12 = 16$ ct. more a dozen; 6 ct. $+$ 16 ct. $= 22$ ct., *Ans.*

(7.) $\frac{1}{8}$ of a dime $= \frac{10}{8} = \frac{5}{4}$ ct.; $\frac{5}{4} \div 3 = \frac{5}{12}$, *Ans.*

(8.) After losing $\frac{3}{8}$, I have $\frac{5}{8}$; spending $\frac{4}{7}$ of this, there are $\frac{3}{7}$ of it left; $\frac{3}{7}$ of $\frac{5}{8} = \frac{15}{56}$, *Ans.*

(9.) A's is $\frac{6}{7}$ as large, and, being $\frac{21}{20}$ as good, is worth $\frac{21}{20}$ of $\frac{6}{7} = \frac{9}{10}$ as much as B's, *Ans.*

(10.) If $\frac{2}{3}$ of A's $= \frac{4}{5}$ of B's, $\frac{1}{3}$ of A's $= \frac{2}{5}$ of B's, and $\frac{3}{3}$ of A's $= \frac{6}{5}$ of B's; hence, $\frac{5}{8}$ of A's $= \frac{5}{8}$ of $\frac{6}{5} = \frac{3}{4}$ of B's, *Ans.*

(11.) After giving $\frac{5}{14}$, there remain $\frac{9}{14}$; B receives $\frac{7}{12}$ of $\frac{9}{14} = \frac{3}{8}$; $\frac{5}{14} = \frac{20}{56}$; $\frac{3}{8} = \frac{21}{56}$; hence, B receives $\frac{21}{56} - \frac{20}{56} = \frac{1}{56}$ of it, more than A, *Ans.*

(12.) B's age is $\frac{5}{3}$ of C's; A's is $\frac{5}{3}$ of B's, and, therefore, $\frac{5}{3}$ of $\frac{5}{3} = \frac{25}{9}$ of C's; $\frac{25}{9} = 2\frac{7}{9}$, *Ans.*

(13.) If $\frac{2}{3}$ of mine $= \frac{4}{5}$ of yours, $\frac{1}{3} = \frac{2}{5}$, and all of mine $= \frac{6}{5}$ of yours; we both have $\frac{6}{5} + \frac{5}{5} = \frac{11}{5}$ of yours; of this $\frac{11}{5}$, $\frac{6}{5}$ will be $\frac{6}{11}$, *Ans.*

(14.) In 1 unit there are 3 thirds; in $\frac{1}{2}$ of a unit there is $\frac{1}{2}$ of 3 thirds $= 1\frac{1}{2}$ thirds, *Ans.*

(15.) $\frac{4}{5} = \frac{4}{5}$ of 3 thirds $= \frac{12}{5}$ of 1 third $= 2\frac{2}{5}$ thirds, *Ans.*; $\frac{5}{6} = \frac{5}{6}$ of 9 ninths $= \frac{45}{6}$ of 1 ninth $= 7\frac{1}{2}$ ninths, *Ans.*; if $\frac{3}{5} = 8$ parts, as shown by the *numerator*, $\frac{1}{5} = \frac{1}{3}$ of $8 = 2\frac{2}{3}$ parts, and $\frac{5}{5} = 5$ times $2\frac{2}{3}$ parts $= 13\frac{1}{3}$ parts in a unit, which will be shown by the *denominator*; hence, $\frac{8}{13\frac{1}{3}}$, *Ans.*

(16.) $\frac{4}{5} - \frac{2}{3} = \frac{12}{15} - \frac{10}{15} = \frac{2}{15}$; $\frac{4}{5} + \frac{2}{15} = \frac{12}{15} + \frac{2}{15} = \frac{14}{15}$, *Ans.*

(17.) $\frac{1}{4} + \frac{1}{5} = \frac{9}{20}$ spent; $1 - \frac{9}{20} = \frac{11}{20}$ left; $\frac{11}{20} - \frac{9}{20} = \frac{1}{10} = \$8$; $\frac{10}{10}$, or money at first, $= \$8 \times 10 = \$80$, *Ans.*

(18.) To be $\frac{7}{5}$ of my present age, I must live $\frac{2}{5}$ as many years as my present age; hence, 12 yr. $= \frac{2}{5}$ of my present age; $\frac{1}{5} = \frac{1}{2}$ of 12 yr. $= 6$ yr., and my present age, or $\frac{5}{5}$, $= 30$ yr., from which $\frac{2}{7}$ must be taken to leave $\frac{5}{7}$ of my present age; $\frac{2}{7}$ of 30 yr. $= 8\frac{4}{7}$ yr., *Ans.*

(19.) Four times $\frac{2}{9} = \frac{8}{9}$, which is $\frac{1}{9}$ less than the number; hence, 12 is $\frac{1}{9}$ of the number, which is $9 \times 12 = 108$, *Ans.*

(20.) $1 - \frac{5}{11} = \frac{6}{11}$; $\frac{2}{3}$ of $\frac{6}{11} = \frac{4}{11}$, to the son; $\frac{6}{11} - \frac{4}{11} = \frac{2}{11}$, to the daughter; $\frac{2}{11} = \$4000$; $\frac{1}{11} = \frac{1}{2}$ of $\$4000 = \$2000$; and property $= 11$ times $\$2000 = \$22000$, *Ans.*

(21.) I sold it for $\frac{5}{4}$ of cost; A, for $\frac{3}{5}$ of its cost to him, $= \frac{3}{5}$ of $\frac{5}{4} = \frac{3}{4}$ of the cost to me; $\$6 = \frac{3}{4}$ of $\$8$. *Ans.* $8.

(22.) B's $= \frac{9}{9}$, A's $= \frac{11}{9}$, both $= \frac{20}{9}$ of B's age; $\frac{20}{9} = 50$ yr.; $\frac{1}{9} = \frac{5}{2}$ yr.; $\frac{11}{9} = 27\frac{1}{2}$ yr., A's age; $\frac{9}{9} = 22\frac{1}{2}$ yr., B's age, *Ans.*

(23.) If $\frac{2}{7}$ are under water, $\frac{5}{7}$ are above; after rising

8 ft. there are $\frac{2}{7}$ above; hence, 8 ft. must be $\frac{5}{7}-\frac{2}{7}=\frac{3}{7}$ of the length; $\frac{3}{7}=8$ ft.; $\frac{1}{7}=\frac{8}{3}$ ft., and $\frac{7}{7}=18\frac{2}{3}$ ft., *Ans.*

(24.) 4 yr. must be the difference between $\frac{3}{4}$ and $\frac{9}{10}$ of B's age; $\frac{9}{10}-\frac{3}{4}=\frac{3}{20}$; $\frac{1}{20}=\frac{1}{3}$ of 4 yr. $=\frac{4}{3}$ yr.; $\frac{20}{20}$ or B's age $=26\frac{2}{3}$ yr.; $\frac{3}{4}$ of $26\frac{2}{3}$ yr. $=20$ yr., A's age, *Ans.*

(25.) The difference between $\frac{3}{4}$ of B's and $\frac{2}{3}$ of B's, must be $5 + $4 = $9; $\frac{3}{4}-\frac{2}{3}=\frac{1}{12}$; if $\frac{1}{12}=$ $9, B's money = 12 × $9 = $108; A's = $\frac{2}{3}$ of $108 + $4 = $76.

Ans. A, $76; B, $108.

(26.) If $\frac{2}{3}$ of A's age $=\frac{3}{4}$ of B's, $\frac{1}{3}$ of A's $=\frac{3}{8}$ of B's, and $\frac{3}{3}$ or A's $=\frac{9}{8}$ of B's; difference, $3\frac{1}{2}$ yr., must be $\frac{1}{8}$ of B's, and B's $=3\frac{1}{2}$ yr. $\times 8=28$ yr.; A's $=28$ yr. $+3\frac{1}{2}$ yr. $=31\frac{1}{2}$ yr.

Ans. A, $31\frac{1}{2}$ yr.; B, 28 yr.

(27.) One boy will do it in 3×7 hr. $=21$ hr., and a man, working $4\frac{1}{2}$ times as fast, in 21 hr. $\div 4\frac{1}{2}=4\frac{2}{3}$ hr., *Ans.*

(28.) 4 more men would make 10 men; 6 men can do it in $5\frac{1}{2}$ da., 1 man, in $5\frac{1}{2}$ da. $\times 6=33$ da., and 10 men in 33 da. $\div 10=3\frac{3}{10}$ da.; time saved $=5\frac{1}{2}$ da. $-3\frac{3}{10}$ da. $=2\frac{1}{5}$ da., *Ans.*

(29.) A man and 2 boys $=5$ boys; if 5 boys do the work in 4 hr., 1 boy would require 4 hr. $\times 5=20$ hr., and 1 man or 3 boys, $\frac{1}{3}$ of 20 hr. $=6\frac{2}{3}$ hr., *Ans.*

(30.) The boy works $9\frac{1}{2}-3\frac{3}{4}=5\frac{3}{4}$ hr., or $2\frac{1}{4}$ hr. less than by the 1st supposition; the man, $9\frac{1}{2}-8=1\frac{1}{2}$ hr. longer; hence, the man does as much in $1\frac{1}{2}$ hr., as the boy in $2\frac{1}{4}$ hr.; which are as 2 to 3; hence, the man does $\frac{3}{5}$, and the boy, $\frac{2}{5}$, in 8 hr.; the man would do the whole in 8 hr. $\div\frac{3}{5}=13\frac{1}{3}$ hr.; the boy, in 8 hr. $\div\frac{2}{5}=20$ hr., *Ans.*

(31.) The 3 men work $4\frac{1}{2}$ da. on what 2 men could have done in the required time. If 3 men work $4\frac{1}{2}$ da., 1 man must work $4\frac{1}{2}$ da. $\times 3 = 13\frac{1}{2}$ da., and 2 men would be employed $\frac{1}{2}$ of $13\frac{1}{2}$ da. $= 6\frac{3}{4}$ da., *Ans.*

(32.) If the work of 9 boys $=$ that of 4 men, the work of 3 boys $=$ that of $\frac{3}{9}$ of 4 men $= 1\frac{1}{3}$ men; 2 men $+ 1\frac{1}{3}$ men $= 3\frac{1}{3}$ men; 3 men work 5 days; 1 man must work 5 da. $\times 3 = 15$ da., and $3\frac{1}{3}$ men, 15 da. $\div 3\frac{1}{3} = 4\frac{1}{2}$ da., *Ans.*

(33.) The man does $\frac{5}{2}$ as much as the boy, or $\frac{5}{2}$ to the boys $\frac{2}{2}$, or 5 parts in 7 parts, or $\frac{5}{7}$; the boy, $\frac{2}{7}$; the first does $\frac{5}{7} - \frac{2}{7} = \frac{3}{7}$ of the work, more than the boy; $\frac{3}{7}$ of 10 A. $= 4\frac{2}{7}$ A., *Ans.*

(34.) One man could do it in $4\frac{1}{3}$ da. $\times 6 = 26$ da., and would do $\frac{1}{26}$ in 1 da.; 6 men would do $\frac{12}{26}$ in 2 da., and there would remain $\frac{14}{26}$ to be done in $1\frac{2}{5}$ da., or $\frac{10}{26}$ in 1 da., which requires 10 men; $10 - 6 = 4$ men., *Ans.*

(35.) There are in all $6\frac{3}{4}$ da. $\times 8 = 54$ da. work. If the 2 men worked from the beginning, the 10 men would do $5\frac{7}{8}$ da. $\times 10 = 58\frac{3}{4}$ da. work; hence, the 2 men must stay away as long after the commencement as it would require them to do $58\frac{3}{4} - 54 = 4\frac{3}{4}$ da. work; $4\frac{3}{4} \div 2 = 2\frac{3}{8}$ da., *Ans.*

(36.) 10 men working constantly, would do $\frac{10}{7}$ as much as 7 men, and, therefore, to do $\frac{1}{7}$ as much, must work $\frac{1}{10}$ of the time, and to do $\frac{7}{7}$, or as much, they must work $\frac{7}{10}$ of the time, and rest $1 - \frac{7}{10} = \frac{3}{10}$, *Ans.*

(37.) 8 men, on coming, must do that part of the work, which 3 men could do in $8\frac{1}{3}$ da., or 1 man in 25 da., and the 8 men in $\frac{25}{8} = 3\frac{1}{8}$ da.; they, therefore, stay away $8\frac{1}{3} - 3\frac{1}{8} = 5\frac{5}{24}$ da., *Ans.*

(38.) They, and those brought with them, must do in 5 da. what 4 men could do in $7\frac{1}{2}$ da., or 1 man in 30 da.; there must, therefore, be $30 \div 5 = 6$ men; 6 men $-$ 4 men $= 2$ men, *Ans.*

(39.) At 6 o'clock the min. hand is 30 min. behind the hr. hand; to be 20 min. behind it must gain $30 - 20 = 10$ min.; to be 20 min. ahead, it must gain $30 + 20 = 50$ min. While the hr. hand passes over 5 spaces, the min. hand traverses 60 spaces, thus gaining 55 min. in 60 min., or 1 min. in $\frac{60}{55} = 1\frac{1}{11}$ min.; to gain 10 min. will require $1\frac{1}{11}$ min. $\times 10 = 10\frac{10}{11}$ min.; to gain 50 min. requires $1\frac{1}{11}$ min. $\times 50 = 54\frac{6}{11}$ min.

Ans. $10\frac{10}{11}$ min. past 6, and $54\frac{6}{11}$ min. past 6.

(40.) If the min. hand is as far *past* 8 as the hr. hand is past 3, it must be 25 min. in advance. At 4 o'clock, it is 20 min. behind, and must gain $20 + 25 = 45$ min., which it will gain in $1\frac{1}{11}$ min. $\times 45 = 49\frac{1}{11}$ min. But if the min. hand is as far *behind* 8 as the hr. hand is in advance of 3, it must be between 6 and 7, and as far behind 7 as the hr. hand is in advance of 4. Hence, while the hr. hand has gone a certain distance, the min. hand, which goes 12 times as fast, must have gone 35 min. (to 7), less that certain distance; therefore, 35 min. less the distance $=$ 12 times the distance, and 35 min. $=$ 13 times the distance; 35 min. $\div 13 = 2\frac{9}{13}$ spaces passed over by hr. hand since 4 o'clock; $2\frac{9}{13} \times 12 = 32\frac{4}{13}$ min. passed over by the min. hand.

Ans. $32\frac{4}{13}$ min. past 4, and $49\frac{1}{11}$ min past 4.

(41.) The hr. hand is a certain space from 5, or 25 min. $+$ that space, from 12; the min. hand is half as far, or $12\frac{1}{2}$ min. $+ \frac{1}{2}$ that space. While the hr. hand moves that space, the min. hand moves $12\frac{1}{2}$ min. $+ \frac{1}{2}$ that space; but while the hr. hand moves 1 space, the min. hand moves 12 spaces; hence, $12\frac{1}{2}$ min. $+ \frac{1}{2}$ that distance $= 12$ distances, or $11\frac{1}{2}$ distances $= 12\frac{1}{2}$ min.; 1 distance $= 12\frac{1}{2}$ min. $\div 11\frac{1}{2} = 1\frac{2}{23}$ min.; 12 spaces $= 1\frac{2}{23}$ min. $\times 12 = 13\frac{1}{23}$ min. Secondly, the hr. hand is 5 min. + a certain distance from 4; therefore, the min. hand must be 10 min. + 2 distances from 4, or 30 min. + 2 distances from 12; hence, while the hr. hand has moved the distance, the minute hand has moved 30 min. + 2 distances; 30 min. + 2 distances = 12 distances, or 10 distances = 30 min., or 1 distance = 3 min. passed over by hr. hand. 3 min. $\times$ 12 = 36 min., passed over by min. hand.

Ans. $13\frac{1}{23}$ min. past 5; 36 min. past 5.

(42.) Of the 8 loaves, A furnishes 5, and eats $2\frac{2}{3}$, giving C $2\frac{1}{3}$ loaves; B furnishes 3, and eats $2\frac{2}{3}$, giving C $\frac{1}{3}$; hence, 8d. must be shared between A and B, in the ratio of $2\frac{1}{3}$ to $\frac{1}{3}$, or 7 to 1; A must have $\frac{7}{8}$ of 8d. = 7d, *Ans.*; B, $\frac{1}{8}$ of 8d. = 1d., *Ans.*

(43.) It goes 1 mi. down in $\frac{1}{15}$ hr., and returns in $\frac{1}{10}$ hr., or 1 mile and back in $\frac{1}{15}$ hr. $+ \frac{1}{10}$ hr. $= \frac{1}{6}$ hr.; 9 hr. $\div \frac{1}{6}$ hr. $= 54$. *Ans.* 54 mi.

(44.) If 15 cows eat as much as 10 horses, 9 cows will eat as much as $\frac{9}{15}$ of 10 horses = 6 horses; hence, I can keep 10 — 6 = 4 horses, *Ans.*

(45.) 12% of B's = 16% of C's; then, 1% of B's = $\frac{4}{3}$% of C's; and B's = $\frac{4}{3}$ of C's; therefore, he has $\frac{1}{3}$ more than C, and \$100 is $\frac{1}{3}$ of C's money; \$100 is $\frac{1}{3}$ of \$300, C's money; 16% of \$300 = \$48, *Ans.*

(46.) If 1 man save $\$1\frac{3}{4}$, 8 men will save 8 times $\$1\frac{3}{4}$ = \$14; \$14 was paid by the 6 new passengers, each paying $\$\frac{14}{6} = \$2\frac{1}{3}$, and 14 passengers pay 14 times $\$2\frac{1}{3} = \$32\frac{2}{3}$, *Ans.*

(47.) There are $\frac{7}{5}$ as many persons; hence, each one must pay $\frac{5}{7}$ as much, thereby saving $\frac{2}{7}$; $\frac{2}{7} = 60$ ct., $\frac{1}{7} =$ 30 ct.; $\frac{7}{7} = 30$ ct. $\times 7 = \$2.10$, *Ans.*

(48.) The 3 lb. worth 4 ct. less per lb. would make a difference of 12 ct. in 8 lb., or $1\frac{1}{2}$ ct. a lb.; $8\frac{1}{2}$ ct. + $1\frac{1}{2}$ ct. = 10 ct.; 10 ct. — 4 ct. = 6 ct.

Ans. 10 ct. and 6 ct. a lb.

(49.) The cost would be 1 ct. less a lb., on 16 lb., that is 16 ct. But the difference is $\frac{1}{3}$ of the cost of 6 lb. of good sugar. 16 ct. is $\frac{1}{3}$ of 48 ct.; 48 ct. ÷ 6 = 8 ct.; $\frac{2}{3}$ of 8 ct. = $5\frac{1}{3}$ ct.; 10 lb. at 8 ct. cost 80 ct.; 6 lb. at $5\frac{1}{3}$ ct. cost 32 ct.; 16 lb. cost 80 ct. + 32 ct. = 112 ct.; 1 lb. cost $\frac{1}{16}$ of 112 ct. = 7 ct.

Ans. Ingredients, 8 ct. and $5\frac{1}{3}$ ct.; mixt. 7 ct. a lb.

(50.) A would have \$30 more than B; he would, also, have $1\frac{1}{3}$ times more; \$30 ÷ $1\frac{1}{3}$ = \$$22\frac{1}{2}$, B would have; \$$22\frac{1}{2}$ + \$10 = $32\frac{1}{2}$, *Ans.*

(51.) Total cost \$1.85, of which B pays 85 ct.; he should pay $\frac{1}{2}$ of \$1.75 = $87\frac{1}{2}$ ct.; $87\frac{1}{2}$ ct. — 85 ct. = $2\frac{1}{2}$ ct. B owes A, *Ans.*

(52.) 285 ft. — 3 ft. = 282 ft.; 282 ft. ÷ 3 ft. = 94 rows; 285 ft. — 6 ft. = 279 ft., length of each row; in stepping from one row to another, he steps 3 ft. between each two rows, in all 279 ft.; 279 ft. × 94 + 279 ft. = 26505 ft. = 5 mi. 6 rd. 6 ft., *Ans.*

(53.) 1 A. = 43560 sq. ft.; 10% of 43560 sq. ft. = 4356 sq. ft.; 43560 sq. ft. — 4356 sq. ft. = 39204 sq. ft.; 39204 ÷ 90 = 435.6 ft. front; \$1000 ÷ 435.6 = \$2.295+, *Ans.*

(54.) \$1 of stock costs 80 ct., and is sold for \$1.10, so that the gain is 30 ct. on 80 ct., or $\frac{30}{80}$ of the cost, = $37\frac{1}{2}$%, *Ans.*

(55.) $100\% - 15\% = 85\%$; 15% of $85\% = 12\frac{3}{4}\%$; $85\% + 12\frac{3}{4}\% = 97\frac{3}{4}\%$; $100\% - 97\frac{3}{4}\% = 2\frac{1}{4}\%$, loss, *Ans.*

(56.) $100\% + 8\% = 108\%$; $12\frac{1}{2}\%$ of $108\% = 13\frac{1}{2}\%$; $108\% + 13\frac{1}{2}\% = 121\frac{1}{2}\%$; 4% of $121\frac{1}{2}\% = 4\frac{43}{50}\%$; $121\frac{1}{2}\% - 4\frac{43}{50}\% = 116\frac{16}{25}\% = 1.1664$; $\$1166.40 \div 1.1664 = \$1000$, *Ans.*

(57.) $\frac{8}{10}\%$ of $\$2500 = \$20$; immediate perpetuity of $20 a year is worth $\$20 \div .06 = \$333.33\frac{1}{3}$, to which add 1st payment, $20, in advance, and we have $\$353.33\frac{1}{3}$; $\$20 \times 12 = \$240$; $\$353.33\frac{1}{3} - \$240 = \$113.33\frac{1}{3}$, gain by the latter, *Ans.*

(58.) Amount of $300 for 1 yr. 10 mo. = $333; $\$1500 - \$333 = \$1167$, the value of the note 1 yr. 10 mo. hence, or 16 mo. after due. Present worth $= \$1167 \div 1.08 = \$1080.56$, *Ans.*

(59.) The true rate is 3%, quarterly. Amount of $1 for 4 periods at 3%, is $1.12550881; $\$1.12550881 - \$1 = .12550881 = 12\frac{550881}{1000000}\%$, *Ans.*

(60.) The length of the cube $= \sqrt[3]{2571353}$ cu. in. = 137 in.; surface of 1 face $= (137 \text{ in.})^2 = 18769$ sq. in., *Ans.*

(61.) The solidity = the cube of the side, = the square of the side multiplied by the side. The surfaces = the square of the side multiplied by 6; if these numbers are equal, the side must equal 6 in., *Ans.*

(62.) $\frac{1}{4}$ of 20 ft. = 5 ft., the side of the square; $(5 \text{ ft})^2 = 25$ sq. ft., the surface of the square; the surface of the circle (Art. 439,) $= (10)^2 \div 3.14159265 = 31.831$ sq. ft. nearly; from this, subtracting 25 sq. ft. gives 6.831 sq. ft. nearly, more in the circle, *Ans.*

(63.) This is the same as an annuity for 20 payments at $2\frac{1}{2}\%$, the present value being \$1200; present value of an immediate perpetuity of \$1 per quarter, at $2\frac{1}{2}\%$, = \$40; present value of same perpetuity, deferred 20 quarters, = \$40 ÷ 1.63861644 = \$24.411; \$40 — \$24.411 = present value of annuity of \$1 = \$15.589; \$1200 ÷ \$15.589 = \$76.98 per quarter = \$307.92 a yr., *Ans.*

(64.) 5% on $\frac{1}{3}$ of \$3000 = \$50; by table (Art. 349), \$50 × 13.390 = \$669.50, *Ans.*

(65.) Present worth of \$200 in 1 mo., at 9%, = \$198.511; of \$200 due in 2 mo., at 9%, = \$197.044; of \$200 due in 3 mo., at 9%, = 195.599; of \$200 due in 4 mo., at 9%, = 194.175; of \$200 due in 5 mo., at 9%, = 192.771: total, \$978.10, *Ans.*

(66.) 100% — 25% = 75%; 25% of 75% = $18\frac{3}{4}\%$; 75% — $18\frac{3}{4}\%$ = $56\frac{1}{4}\%$ = \$675; 1% = \$12, and 100% = \$1200, *Ans.*

(67.) Int. of \$1 for 63 da. = \$.021; the true discount would be $\frac{.021}{1.021} = \frac{21}{1021}$; loss $= \frac{21}{1000} - \frac{21}{1021} = \frac{441}{1021000}$; if $\frac{441}{1021000}$ of the principal = \$4.80, the principal must be \$4.80 ÷ 441 × 1021000 = \$11112.93, *Ans.*

(68.) I lost 20% of \$10000 = \$2000, and have left \$8000, on which 18% is \$1440; \$2000 — \$1440 = \$560, to be gained on money borrowed; 18% — 4% = 14% gain; \$560 is 14% of \$560 ÷ .14 = \$4000, *Ans.*

(69.) The quantity will be equal to the solidity of a body 38 ft. by 52 ft., and $\frac{1}{4}$ in. thick; 456 in. × 624 in. × $\frac{1}{4}$ in. = 71136 cu. in.; 71136 cu. in. ÷ 231 = 307.948 gal., which ÷ $31\frac{1}{2}$ = 9.776+ bbl., *Ans.*

(70.) The sums which would amount to \$1 in 15, 13,

11, and 9 yr. respectively, are found by rule (Art. 338), and are \$.51672044, \$.56427164, \$.61619873, and \$.67290442 whose sum is \$2.37009523:

\$2.37009523 : \$.51672044 : : \$20000 : \$4360.34, 1st *Ans.;*
\$2.37009523 : \$.56427164 : : \$20000 : \$4761.59, 2d *Ans.;*
\$2.37009523 : \$.61619873 : : \$20000 : \$5199.78, 3d *Ans.;*
\$2.37009523 : \$.67290442 : : \$20000 : \$5678.29, 4th *Ans.*

(71.) $3\frac{1}{2}\%$ of \$30000 = \$1050, semi-annual payment; \$1050 ÷ .04 = \$26250, the present value of the payments, if that amount is to be paid at the end of the time; but \$30000 — \$26250 = \$3750; hence, if \$26250 were paid, there would be a gain, at the end of 20 yr., of \$3750, the present value of which is \$3750 ÷ 4.80102063 = \$781.08; \$26250 + \$781.08 = \$27031.08, *Ans.*

(72.) First, let us suppose there are only as many different kinds of animals as there are prices, that is, three kinds. Performing the work according to the

	9	5	3	2	29	30	31	32	33	34	35	36	37
4	2	2		5	68	60	52	44	36	28	20	12	4
	1	3	5		3	10	17	24	31	38	45	52	59
			8	7									

method laid down under the article of Alligation, we find by taking 5ths of 8 and 5ths of 7 to make 100, we have *nine* different answers. But in *each* answer *required* by the dealer, there must be *two* kinds for the *one price* \$9. Consider the first one 29. In this there could be:

28 hogs and 1 calf,
27 " " 2 calves,
26 " " 3 calves,

and so on, to 1 hog and 28 calves; that is, corresponding to the first column above, there could be 28, (or 29 — 1) answers ; so, corresponding to the next there could be 29 answers. In all, we could have answers,

$$28 + 29 + 30 + 31 + 32 + 33 + 34 + 35 + 36 =$$

288 answers, *Ans.*

(73.) A's gain is $\frac{112}{350} = 32\%$; B's $\frac{88}{220} = 40\%$; C's $= \frac{120}{350}$, or 48% ; C gains 8% more than B, by contributing his stock 2 mon. longer ; hence, the gain is 4% a mon. ; $32\% \div 4\% = 8$ mon. A's, *Ans.* ; $40\% \div 4\% = 10$ mon. B's, *Ans.* ; $48\% \div 4\% = 12$ mon., C's, *Ans.*

(74.) The discount is 20% of the face, or $22\frac{1}{2}\%$ of the proceeds ; hence the proceeds $= \frac{20}{22\frac{1}{2}} = \frac{8}{9}$ of the face, and the discount $= \frac{1}{9}$ of the face $= 11\frac{1}{9}\%$; time $= \frac{11\frac{1}{9}}{20}$ yr. $= \frac{5}{9}$ yr. $= 200$ da., *Ans.*

(75.) A receives \$57.90 — \$29.70, = \$28.20, more than B ; since he contributed $\$7.83\frac{1}{3}$ more than B, his investment must have been $\frac{7.83\frac{1}{3}}{28.20} = \frac{5}{18}$ of his gain ; $\frac{5}{18}$ of \$57.90 $= \$16.08\frac{1}{3}$, A, *Ans.* ; $\frac{5}{18}$ of \$29.70 = \$8.25, B, *Ans.*

(76.) Suppose he borrows \$1 ; at the end of 6 mon. it amounts to \$1.0609, of which he pays 3 ct. for interest, leaving \$1.0309, which, in the next 6 mon., will amount to \$1.09368181, from which, on paying 3 ct. int., he will have remaining \$1.06368181, thus clearing, in the year, \$.06368181 on each \$1 borrowed ; $\$2450.85 \div .06368181 =$ \$38485.87, *Ans.*

(77.) Discount for 4 yr. at $4\% = \frac{.16}{1.16} = \frac{4}{29}$ of debt ; for 4 yr. at 6%, it is $\frac{.24}{1.24} = \frac{6}{31}$ of debt ; $\frac{6}{31} - \frac{4}{29} = \frac{50}{899}$; $\frac{50}{899}$ of debt = \$25 ; $\frac{1}{899} = \$\frac{1}{2}$; and the debt is \$449.50, *Ans.*

(78.) \$1 — 25 ct. = 75 ct.; \$2700 ÷ .75 = \$3600; 8% of \$3600 = \$288, income; \$288 ÷ .10 = \$2880 par value; \$2880 × .96 = \$2764.80, *Ans.*

(79.) 69% is $\frac{3}{4}$ of 92%; I will, therefore, receive $\frac{3}{4}$ as much stock at 92%, as at 69%; 7% on $\frac{3}{4}$, is the same as $5\frac{1}{4}$% on the whole; so that I gain $5\frac{1}{4}$% — 5% = $\frac{1}{4}$%, annually, on \$5200; $\frac{1}{4}$% of \$5200 = \$13; \$13 a year is worth \$13 ÷ .06 = \$216.66$\frac{2}{3}$, *Ans.*

(80.) $2\frac{1}{2}$% of \$1500 = \$37.50; \$1500 — \$37.50 = \$1462.50 received; \$1462.50 ÷ 1.15 = \$1271.739, the cost with int. for 3 mon.; am't of \$1 for 3 mon. at 6% = \$1.015; \$1271.739 ÷ \$1.015 = \$1252.94, *Ans.*

(81.) The 9th term must be multiplied by the ratio 4 times, to give the 13th term; hence, 11160261 ÷ 137781 = 81, is the 4th power of the ratio; $\sqrt[4]{81} = \sqrt{9} = 3$; the 9th term must be divided by the ratio 5 times, or by the 5th power of the ratio, to give the 4th term; $(3)^5 =$ 243; 137781 ÷ 243 = 567, *Ans.*

(82.) \$19487.171 ÷ \$13310 = 1.4641, ratio of increase for 4 yr.; by reference to the table (Art. 348), this is found opposite 4 yr., under 10%, the required rate; am't of \$1 for 3 yr., at 10% = \$1.331; \$13310 ÷ 1.331 = \$10000. *Ans.* Capital, \$10000; rate, 10%.

(83.) 3 min. 28 sec. : 60 min. : : 4 in. : 69.23077 in., circumference; 69.23077 in. ÷ 3.14159265 ÷ 2 = 11.02 — in., *Ans.*

(84.) Radius of whole circle = 16 in.; radius of aperture = 3 in.; area of stone face = 256 × 3.1416 — 9 × 3.1416 = 247 × 3.1416. Each is to have $\frac{247}{3}$ × 3.1416;

and hence this amount + aperture = circle left by second man; *twice* that amount + aperture = circle first left. Thus:

$$(3^2 + \tfrac{247}{3}) \times 3.1416 = \text{2d circle left.}$$
$$(3^2 + \tfrac{494}{3}) \times 3.1416 = \text{1st} \quad " \quad "$$

But if we divide the area of any circle by 3.1416, the result is the square of the radius. Hence, $\sqrt{3^2 + \frac{247}{3}} =$ radius 2d left $= 9.557-$; $9.557 - 3. = 6.557-$, *Ans.* $\sqrt{3^2 + \frac{494}{3}} = 13.178$, the 1st rad. left; $13.178 - 9.557 = 3.621$, *Ans.;* $16. - 13.178 = 2.822$, *Ans.*

(85.) The supposed cost = the actual cost + \$300; 20% or $\frac{1}{5}$ of the supposed cost = $\frac{1}{5}$ of the actual cost + \$60; the difference between gaining $\frac{1}{5}$ of the cost and losing $\frac{1}{5}$ of the cost + \$60, is $\frac{2}{5}$ of the cost + \$60; hence, $\frac{2}{5}$ of the cost + \$60 = \$300; $\frac{2}{5}$ of the cost = \$240; $\frac{1}{5}$ is \$120, and the cost = \$120 × 5 = \$600, *Ans.*

(86.) It goes down 1 mi. in $\frac{1}{16\frac{1}{4}} = \frac{4}{65}$ hr., and up 1 mi. in $\frac{1}{10}$ hr.; $\frac{1}{10} - \frac{4}{65} = \frac{1}{26}$ hr. longer in going up 1 mi. than in going down the same distance; $22\frac{1}{2} \div \frac{1}{26} = 585$ mi., *Ans.*

(87.) The actual cost was $\frac{10}{9}$ of the supposed cost; therefore, the selling rate per cent was $\frac{9}{10}$ as much, and the difference between the two selling rates is $\frac{1}{10}$ of the supposed selling rate, which is $\frac{1}{9}$ of $\frac{9}{10}$; hence, 15% is $\frac{1}{9}$ of the actual selling rate; 15% × 9 = 135%, selling rate; 135% − 100% = 35%, *Ans.*

(88.) Each \$1 in the face of the check cost me 55 ct.; I receive for it \$1 ÷ .60 = \$$1\frac{2}{3}$ in bonds; 7% of \$$1\frac{2}{3}$ = $11\frac{2}{3}$ ct.; hence, I receive yearly $11\frac{2}{3}$ ct. on 55 ct., which is $21\frac{7}{33}$%, *Ans.*

(89.) \$1 — 6 ct. — 30 ct. = 64 ct. left in sugar for each \$1 invested. The molasses losing 40%, brought 60% of 30 ct., or 18 ct., making in all 82 ct. for each \$1 invested. The sale must be \$1.14, an increase of 32 ct., and as the 64 ct. must yield this, the % is 32 ÷ 64 = 50%, *Ans.*

(90.) Suppose each man can remove 10 parts in a minute. [Whatever he removes can be considered as of 10, or any other number of parts.] Then, by the conditions, we have the statements:

6 men remove 60 parts in 1 minute.
6 " " 3600 " " 60 minutes.
1 " " 2200 " " 20 "

In either of the two cases the dock is cleared; and in the 60 minutes the dock has had 1400 more of parts to be removed than in the 20 minutes. As the amount *on* the dock *at first*, is the same in either case, the *difference in amount*, 1400 parts, must be what would run on in the *difference of time*, 40 minutes; hence, 35 parts run on in 1 minute. Then, since at this rate, 2100 parts run on in an hour, and *in* that hour all the parts removed are 3600, it follows that 1500 parts *are on* the dock when the work begins. Now, since 35 parts run on in 1 minute, it will take $3\frac{1}{2}$ times the work of 1 man to *keep clearing away* the *supply* of each minute; consequently, four men being the given force, there is left *half* the work of one man, each minute, toward clearing the original amount of 1500 parts; but half a minute's work for one man, removes 5 parts; hence, it will require, to remove the whole, 1500 ÷ 5, or 300 minutes; that is, 5 hr., *Ans.*

(91.) For each dollar in the worth of the apples he received in the sale 96 ct. Of this 96 ct. there were two parts; one, an investment, and the other, 8% of the investment; hence, the investment was $\frac{100}{108}$ of the 96 ct., or $88\frac{8}{9}$ ct. Had no money been reserved, the loss would have been $11\frac{1}{9}$ ct. on each \$1 of the original value; that is, he would have lost $\frac{1}{9}$ of the whole had he left the \$18 in the proceeds. But the loss on that \$18 would have been simply the commission, $\frac{8}{108}$ of \$18, or $\$1\frac{1}{3}$. Hence, had he kept no money, the whole loss would have been $\$32 + \$1\frac{1}{3} = \$33\frac{1}{3}$; and as this would have been $\frac{1}{9}$ of the value of the apples, that must have been 9 times $\$33\frac{1}{3}$, or \$300, *Ans.*

(92.) 1. The gain being $3\frac{1}{2}$ minutes each day, the clock indicates a period of 1443.5 when the true period is 1440. Hence, since on the 29th the indicated period was 7 days,

1443.5 min. : 1440 min. : : 7 days : true lapse.
$7 \times 1440 \div 1443.5 = 6.98302$ da. =
6 da. 23 hr. 35 min. 33.5 sec.; hence,
35 min. 33.5 sec. past 11 A. M., 29th, *Ans.*

2. To show the same time on the clock face, it must gain a whole *half-day*, or, 720 min.; hence,

gain in 1 day : time required : : req. gain : req. time.
3.5 min. : 1440 min. : : 720 min. : ?

$$\frac{720 \times 1440}{3.5} \text{ min.} = \frac{1440}{7} \text{ da.} = 205\tfrac{5}{7} \text{ da.}$$

$205\frac{5}{7}$ da. after Feb. 22d (noon), leap-year, is $\frac{5}{7}$ da. after the noon of Sept. 14th; hence,

Sept. 15th, $8\frac{4}{7}$ min. past 5 A. M., *Ans.*

(93.) Since 144 = base $\times$ $\frac{1}{2}$ height, or, in *this* case, base $\times$ base, the *area is the square of* $\frac{1}{2}$ *the height.* And as the triangles formed are *similar*, we have $6^2 = 36$ sq. in., and so, by squaring half each alt., $42\frac{1}{4}$ sq. in., 49 sq. in., $52\frac{9}{16}$ sq. in., *Ans.*

(94.) 1. The case is one of Arith. Prog. (Art. 393), and as first term is 16, second 48, com. diff. 32, the 5th term is

$$16 + (5 - 1) \times 32 = 144.$$

$$\text{Sum} = \frac{(144 + 16)}{2} \times 5 = 400 \text{ ft., } Ans.$$

2. Observe that it *constantly* increases; in the latter half of any period, falling farther than in the first half. Hence, in the latter half of the 5th second it falls *more* than the *half* of 144 ft. The figure will illustrate this. The length being twice the breadth, in any one of the triangles, the addition made for each step from the vertex toward the base is a trapezoid, and the lower half of the altitude, as KN, in any case, bounds a trapezoid smaller than OM, but greater than its half. The whole distances are as the squares of the times, as the *whole areas* are proportional to the squares of the altitudes, in the former problem.

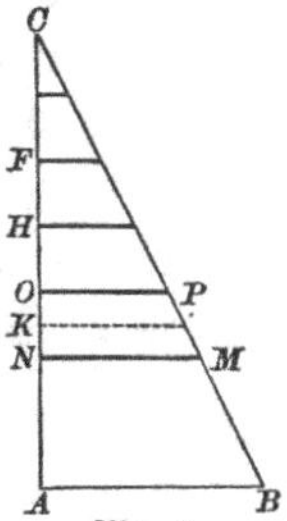

Fig. 5,

16 ft. $\times (4\frac{1}{2})^2 = 324$ ft., *Ans.*; so $4^2 \times 16$ ft. $= 256$ ft., *Ans.*

REMARK.—Compound interest presents a case of *constant* increase, in some respects like these.

(95.) As in the two preceding problems, the distances are as the squares of the times; hence, $6\frac{1}{2} \times 6\frac{1}{2} = 42\frac{1}{4}$ miles, *Ans.*

(96.) The side being to the front as 2 to 3, the whole body is now $\frac{3}{2}$ of a square, whose side is the present width. The increase will be 16 men at the corner, 4 ranks at the side, and 4 ranks at the end. The former 4 being $\frac{3}{2}$ of the latter 4, the side rank *added* to an end rank will make 10 ranks of the same length as the present width. Hence, without the 16 at the corner, the new body will consist of 2304 men, and will equal $\frac{3}{2}$ of a square body of the present width, + 10 ranks of that length.

For convenience, let each man occupy the space of a square yard. Then, $\frac{3}{2}$ of a square + 10 strips of the same length, 1 yard wide, will make 2304 sq. yd.; or, taking $\frac{2}{3}$ of this space, *one* such square + $\frac{20}{3}$ of such a strip $= 1536$ sq. yd.; that is, 1 such a square and *two* strips $\frac{10}{3}$ yd. wide, make 1536 sq. yd. If these strips be put on two adjacent sides of the square, we shall require yet a corner square of $\frac{100}{9}$ sq. yd. to make one full square $= 1536 + \frac{100}{9} = \frac{13924}{9}$ sq. yd.; the side of this $=$ $\frac{118}{3}$ yd.; this being $\frac{10}{3}$ yd. longer than the present width, the required width $= 36$ yd. Hence, there are 36 men in an end rank, 54 in a side rank, and in all $54 \times 36 = 1944$ men, *Ans.*

(97.) If he had cut off four such pieces, he would have left a square piece of cheese, the largest possible. Hence, the four segments weighing 12 lb., the whole cheese is to 12 lb., as a circle to the four segments left by cutting out the largest square. The diagonal of that square will be the diameter, and it will be a hypothenuse in a right-angled triangle of equal base and perpendicular; hence, for each one in the diagonal there are $\sqrt{\frac{1}{2} \text{ of } 1}$, or $\sqrt{.5}$ in the side. Hence, for each 1 inch in the diagonal, there is $(\sqrt{.5})^2$, or $\frac{1}{2}$ square inch in the square; but for each 1 inch in the diameter, there are $\frac{3.1415926}{4}$, or .785398 sq. in. in the circle, and the four segments around the square have .785398 — .5 = .258398. Therefore, we have,

4 segments : circle :: 12 lb. : whole cheese;
.258398 : .785398 :: 12 lb. : 33.0232 lb., *Ans.*

(98.) Let P be the highest and K the lowest point of the wheel; let AB represent the mud line across the wheel. OP = 2; ON = 1, because NK = 1. ANO is a right-angled triangle, having hypothenuse = 2, perpendicular = 1, and hence the base AN = $\sqrt{4-1} = \sqrt{3}$, and the whole mud line AB = $2\sqrt{3}$. In the right-angled triangle NPA, the base = $\sqrt{3}$, the height 3, and the hypothenuse PA = $\sqrt{3+9} = \sqrt{4 \times 3} = 2\sqrt{3}$. But this is the same value which was found for AB; and as the same can be found for BP, it follows that the mud line marks off one of the three segments left by taking the largest equilateral triangle out of the circle. The area of the triangle = $\frac{1}{2}$ of BA × PN = $\sqrt{3} \times 3$ = 5.1961524. The whole circle = 4^2 × .785398 = 12.566368. Hence, the lower part marked off = $\frac{1}{3}$ of (12.566368 — 5.1961524) = 2.456738. Twice this segment + the triangle = 10.109628 = the part of the whole circle above the line. The fraction of the wheel which is out of the mud, is the same fraction which 10.109628 is of the circle area 12.566368, or

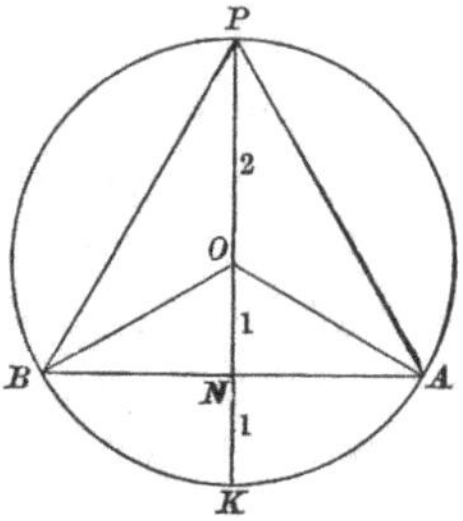

Fig. 6.

$$\frac{10.109628}{12.566368} = .804498+, \textit{ Ans.}$$

(99.) The length being $\frac{5}{3}$ of the breadth, the lot is $\frac{5}{3}$ of the square of the breadth; hence, $\frac{1}{3}$ of the square of the breadth is 27 sq. rd., and the whole square of the breadth = 81 sq. rd.; therefore, the breadth is 9 rd. and the length 15 rd. Suppose the road marked out, and suppose also that we had four such lots disposed in the form of a square (as the pupil can arrange four Eclectic Third Readers in the form of a square inclosing a hollow square), the *end* of each, touching the longer side of the next; the *side* of the whole square would be equal to the sum of the length and breadth of each piece,—in this case, 15 rd. + 9 rd., or 24 rd. The required *road* in *one*, joining the *like* road in another, there will be a square frame around the whole four. The area of the large square is 576 sq. rd.; *one* such road being $\frac{1}{4}$ of 135 sq. rd., the whole road-frame contains 135 sq. rd., and incloses 441 sq. rd., a square whose side is 21 rd. This differs from 24 rd. by twice the width of the road, and therefore that width $= \frac{1}{2}$ of $(24 - 21)$ rd. $= 1\frac{1}{2}$ rd. $= 24\frac{3}{4}$ ft., *Ans.*

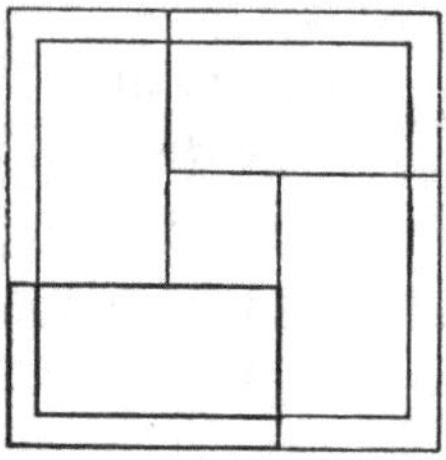

Fig. 7.

(100.) Three equal circles, of a radius $= 1$, will be to a large circle just touching them as the required grass areas to the given lot. Let those of the figure be three such circles, and let their centers be joined, forming an equilateral triangle, whose side $= 2$. From O, the lines to F, H, and N divide the triangle into 3 equal parts; that is, HON is $\frac{1}{3}$ of HFN; having the *same base*, its altitude, therefore, must be $\frac{1}{3}$ of FA. Hence, FN being 2, and AN, 1, FA $= \sqrt{4 - 1} = \sqrt{3} = 1.7320508075$, and $\frac{2}{3}$ of this $=$ FO; also $\frac{2}{3}\sqrt{3} + 1$, or OF + FP $=$ OP $= 2.1547005383$. Now, by proportion, the given radius being $7\frac{1}{2}$ rd.,

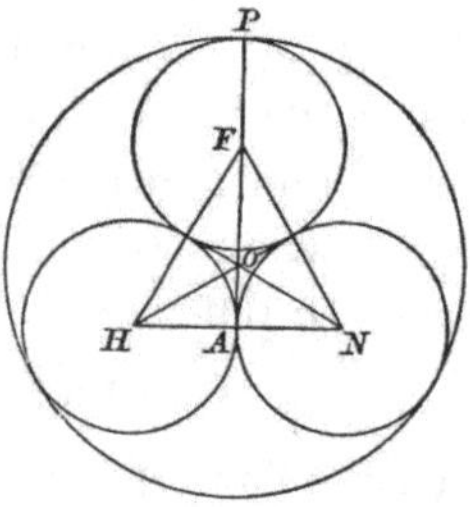

Fig. 8.

2.1547005383 : 1 :: 7.5 rd. : 3.4807621125 rd.
And, 3.4807621125 rd. $\times$ 2 $=$ 6.9615242$+$ rd., distance between their centers, *Ans.*

Again: the triangle [Art. 385, 4, Rem.] is composed of the triangular space within $+$ $\frac{1}{6}$ of each small circle; hence, FHN$-\frac{1}{2}$ a small circle$=$the triangular space. One small circle (rad. 3.4807621125)$=$3.14159265358$\times$3.4807621125^2 $=$ 12.1157048838 $\times$ 3.14159265358 $=$ 38.0626094569; $\frac{1}{2}$ of this$=$19.0313047$+$.

The side of the equilateral triangle is 6.961524225, and the half base $=$ 3.4807621125.

Perp. $= \sqrt{48.46281953 - 12.11570488} = 6.028856828.$
Area $=$ 6.0288568 $\times$ 3.48076211 $=$ 20.9850162$+$.
20.985016 $-$ 19.031304 $=$ 1.953712$-$, *Ans.*

(101.) Consider this board as a trapezoidal piece cut off from a triangle. The area is $(7+17)\times 30 = 720$ sq. in. The required area to be cut off below, is 360 sq. in. Let the figure ABCD represent the face of the board, APB the triangle from which it was cut. The whole triangle is similar to HBC or to the piece PDC. [See Art. 389, Rem. 2, and Art. 231, Prob. 8.] Hence, as the piece HBC has a base 17 $-$ 7, or 10, and a height 6 times this, so the whole triangle has a height of 6 times 17, or 102 inches; the area of it $=\frac{1}{2}$ of $102\times 17 = 867$ sq. in. Now, the problem is the same as to cut off 360 sq. in. from the triangle, by a line such as ON; *and the triangle left* will have 507 sq. in. Hence, the height $\times$ $\frac{1}{2}$ the base $=$ 507; or, the height $\times$ the whole base $=$ 1014. But the base is $\frac{1}{6}$ of the height; hence, the height $\times$ by $\frac{1}{6}$ of itself makes 1014, and the height $\times$ by the *whole* of itself makes 6084;

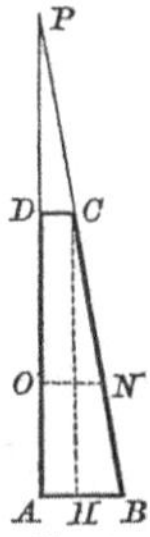

Fig. 9.

hence, the height $= \sqrt{6084} = 78$ in. Then, PA — PO, or $102 - 78 = 24$ in. $= 2$ ft., *Ans.*

(102.) Four equal circles, of radius 1, will be to a large circle inclosing and touching them as the required pieces to the given plate. Let DO, in the figure, be 1; then OC $= \sqrt{2}$, and CP $= 2.4142135624$. Area of the large circle is $2.4142135624^2 \times 3.1415926535$. That of small one is 3.1415926535; hence, large circle is to small one, as $2.4142135624^2 : 1$, and

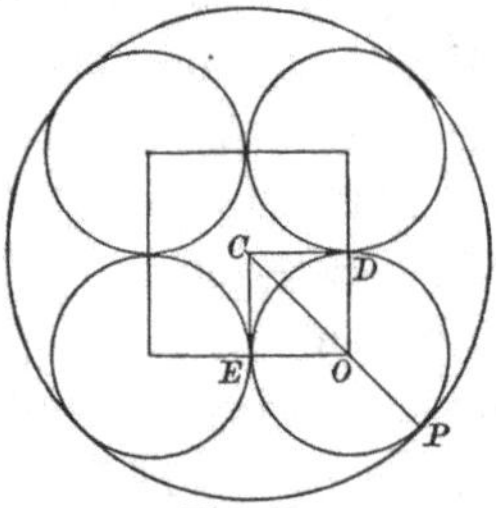

Fig. 10.

$2.4142135624^2 : 1 :: \$67$: worth of 1 circle piece;
$5.8284271248 \ \ : 1 :: \$67 : \$11.4953826384$, *Ans.*

The square, formed by joining the centers, includes $\frac{1}{4}$ of each circle; hence, the *parts* of the *small* circles, *around* the square, make *three whole circles.* Hence, if the area of the square and 3 times the area of one circle be taken from the large circle, the remainder is the area of the four outer portions.

Area of the whole figure $= 2.4142135624^2 \times 3.1415926535 = 18.31054383724$; the square $+ 3$ circles $= 4 + 3$ times $3.1415926535 = 13.4247779605$; this, taken from large circle, leaves 4.88576587674. Then,
$18.31054383724 : 4.88576587674 :: \$67 : \$17.87747631$, *Ans.*

(103.) The diameter of the given ball is to the distance of its extreme point from the corner, as the required smaller diameter is to the distance of the nearest point of the given ball from the corner.

If the 12-inch ball were cut out of a cube, the length of the diagonal cut off at each end to make the diameter, would be the same as the distance from nearest point of ball to the corner. The diameter would be the length of edge of the cube; $\sqrt{288}$, the diagonal of a *face*, and $\sqrt{288+144} = 12\sqrt{3} = 20.7846096$, the diagonal of the cube. $\frac{1}{2}$ of $(20.7846096 - 12) = 4.3923048$, the distance from given ball to corner. The distance from extreme point to corner $= 16.3923048$, and, by statement above,

16.3923048 in. : 12 in. : : 4.3923048 in. : 3.2154 in., *Ans.*
(The exact quotient is 3.2153904).

In like manner this smaller diameter is to the given one as the given one to a required larger one standing in the same relation;

$$3.2154- : 12 :: 12 : \frac{144}{3.2154-} = 44.7846, \textit{ Ans.}$$

(104.) The log was equal to two cubes. Suppose, then, the trough divided into two equal parts by a cut parallel to the ends, and let one half of it be represented by the figure. That will contain 5886 cu. in. of wood. The student has learned that a cubical box, of sides 3 inches thick, may be considered as made up of 6 square blocks, 12 blocks of the same length, 3 inches wide, and 8 corner cubes, of 27 cu. in each. The half-trough, represented, is exactly what such a box would become, if two of the square blocks and 1 long one were taken out. Hence, the 5886 cu. in. may be considered as made of 4 square blocks, 11 long blocks, and 216 cu. in.; that is, 4 square blocks and 11 of the same length, 3 inches wide, make 5670 cu. in.; and, taking a *side surface* in each, we have 4 squares and 11 strips = 1890 sq. in. *One* square + $\frac{11}{4}$ strip, 3 in. wide, =

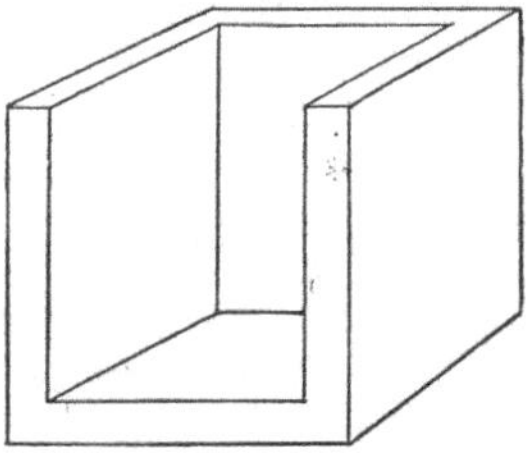

Fig. 11.

472.5 sq. in. If the strip $\frac{11}{4}$ of 3 in. wide, be divided into two, each $\frac{33}{8}$ in. wide, and placed on adjacent sides of the square, there will be wanting a square of $(\frac{33}{8})^2$, or $\frac{1089}{64}$ sq. in., at the corner to make a complete square of 489.515625 sq. in.; the side of this is $22\frac{1}{8}$ in., and, taking away the added $\frac{33}{8}$ in., there are left 18 in., which are 6 in. less than the thickness of the log. Hence, the capacity =

$(2 \times 24 - 6) \times (24 - 6) \times (24 - 3) \div 231 = 68\frac{8}{11}$ gal., *Ans.*

(105.) The whole vessel may be considered as a frustum cut off from a cone, a section of which is shown in the figure.

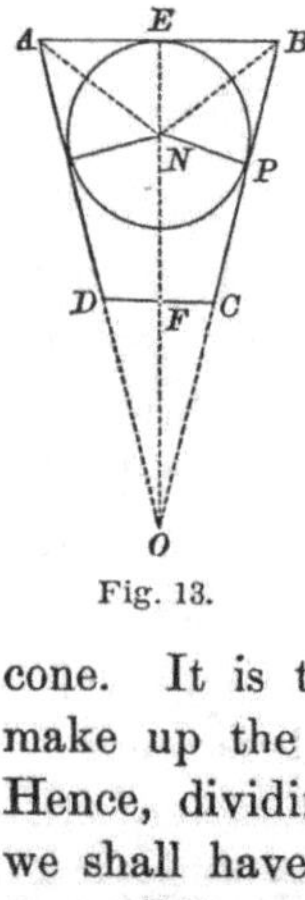

Fig. 13.

AB = 9, DC = $4\frac{1}{2}$, EF = 10, and by prop., EO being twice EF, EO = 20, and so find OB = $20\frac{1}{2}$. The mouth of the vessel, or base of the cone, $= 9^2 \times .7854$, and the whole cone would hold $81 \times \frac{1}{4} \times 3.1416 \times \frac{20}{3} = 135 \times 3.1416$. The small cone cut off being $\frac{1}{8}$ of the large one, the capacity of the vessel is $\frac{7}{8}$ of the large cone, or $\frac{945}{8} \times 3.1416$. Hence, the water in the vessel is $\frac{945}{32} \times 3.1416$. Let NP or NE be the radius of the largest ball which could be put in the cone. It is the altitude of each of three triangles which make up the triangle ABO, whose area $= 4\frac{1}{2} \times 20 = 90$. Hence, dividing double of this by the sum of the bases, we shall have $180 \div (41 + 9) = 3.6$ the radius NP; and the solidity of that largest ball $= \frac{7776}{125} \times 3.1416$. Now, the water in the vessel, + a cone DCO, would make a cone $(\frac{945}{32} + \frac{135}{8}) \times 3.1416$; and the problem is the same as to find a ball which could be put in a cone and be just covered by this amount of water, *if* such ball do *not* extend below DC. But, the cones being similar, the *water* which would cover the *largest* ball is to the *water* covering the *required* ball as the largest ball is to the required ball; the water covering the largest is $(135 - \frac{7776}{125}) \times 3.1416$; hence, $\frac{9099}{125} \times$

$3.1416 : \frac{1485}{32} \times 3.1416 : :$ cube of large diameter : cube of required one [Art. 389, 4.]

$\frac{337}{125} : \frac{55}{32} : : (\frac{36}{5})^3 : (\frac{36}{5})^3 \times (\frac{5}{2})^3 \times \frac{55}{1348}$, the cube of req. diam., which therefore equals $\sqrt[3]{5832 \times 55 \div 1348} =$, in inches, $18 \times .3442634 = 6.1967 +$ inches, *Ans.*

(106.) If the box were but 1 in. deep it manifestly would hold at least 50 balls.

These would just fill the box if arranged in 10 rows of 5 balls each, the centers of adjacent rows being 1 in. apart.

But if the balls in the *even* rows (2d, 4th, 6th, &c.) are placed in the cavities between balls in the odd rows (1st, 3d, 5th, &c.) the centers of adjacent rows will be only .86603 —

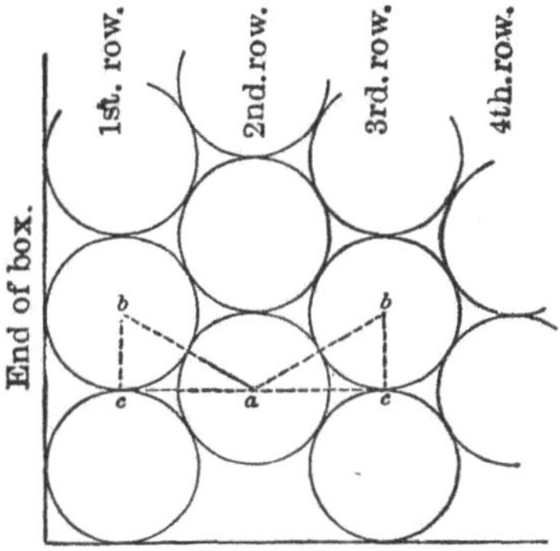

$ab = 1$ in.

$cb = \frac{1}{2}$ in.

$\sqrt{1^2 - \frac{1}{2}^2} = .86603 -$

Side of box.

Fig. 1. Plan of 1st layer.

in. apart and 11 rows can be placed in the box, the odd rows having 5, and the even rows 4 balls, and an unoccupied space of .339 + in. will be left between the outside of the 11th row and the end of the box; that is, the 50 balls so arranged occupy a space 1 in. high, 5 in. wide, and 9.661 in. long, their aggregate volume filling $\frac{54}{100}$th of this space. This is evidently the most compact arrangement possible for a single layer of balls.

Now as to the most compact arrangement for two or more layers. If a ball were placed on top of each ball in the 1st layer, the two layers would contain 100 balls, and would occupy a space $2 \times 5 \times 9.661$ in.; the aggregate volume of balls filling $\frac{54}{100}$th of the total volume of space as before. If, however, the balls in the 2nd layer are placed in the trihedral pits between the adjacent rows in the 1st layer, the plane passing through the centers of balls in the 2nd layer will be but .8165 in. above a similar plane in the 1st layer, and the two layers will go into a box 1.8165 in. high.

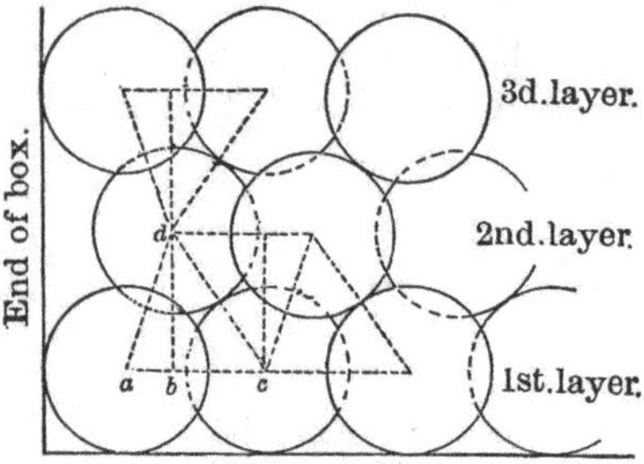

Bottom of box.
Fig. 2.

$cd = 1$ in.

$ac = ad = .86603 -$ in.

Area $acd = \frac{1}{2} cd \sqrt{ad^2 - \frac{1}{4} cd^2}$
$= .35355 +$

By rule II. p. 383.

$$\frac{.35355 +}{\frac{1}{2} ac} = db = .8165 -$$

also,

$$\sqrt{ad^2 - db^2} = ab = .289 -$$

Since the centers of the rows in the 2nd layer are .289 in. to the right of the center of the corresponding row in the 1st layer, and since the outside of the 11th row in the 1st layer is .339 in from the end of the box, there is $.339 - .289 = .05$ in. more than enough room in the box for 11 rows in the 2nd layer; that is, the two layers occupy a space 1.8165 in. high, 5 in. wide, and $10 - .05 = 9.95$ in. long. But the 2nd layer contains only 49 balls—the six odd rows containing 4, and the five even rows, 5 balls each; therefore the aggregate volume of the 99 balls in the two layers fills more than $\frac{57}{100}$th of the volume of the space occupied. Hence, this is the most compact possible arrangement of layers, the loss of one ball in the even layers being more than compensated by the gain in space, caused by decreasing the height between all the layers to .8165 in.

Since the depth of the box is 10 in., it will contain 12 layers $\left(\frac{10-1}{.8165}+1\right)$ leaving .0185 in. to spare above the upper layer.

The 6 odd layers contain	50	balls	each	= 300
" 6 even " "	49	"	"	= 294
Total number of balls in box,				594

The aggregate volume of the 594 balls fills $\frac{622}{1000}$th of the total volume of the box. The same number of balls could be placed in a box 5 × 9.95 × 9.9815 in., and would occupy $\frac{626}{1000}$th of its volume.

The answer was obtained by MR. A. J. TRAPP, Pleasant Hill, Mo.

Who Is Joseph Ray?

Joseph Ray lived as a contemporary of Abraham Lincoln. Youth in that generation finished their schoolbooks and then read the Bible, sang from the hymnbooks of Lowell Mason, and read Roman and Greek classics in the original languages. It was not unusual for a blacksmith to carry a Greek New Testament under his cap for reading during his lunch break. The literacy rate, even on the frontier, was higher than today's rate.

Ray was Professor of Mathematics for twenty-five years at a preparatory school in Ohio. He had no use for indolence and sham. He was always delighted to join his students in sports. He knew how to use balls, marbles, and tops as concrete illustrations to help young children make the transfer from solid objects to abstract figures.

From the Presidency of Abraham Lincoln to that of Teddy Roosevelt, few Americans went to school or were taught at home without considerable exposure to either Ray's Arithmetics or McGuffey's Readers—usually both. Ray and McGuffey challenged students to excellent accomplishment. Their influence on our country has certainly eclipsed Mann's and rivaled Dewey's, yet education histories, edited by humanists, seldom mention these men.

Ray's classic Arithmetics are now brought to a new generation which is in search of excellence.

Ray's Arithmetic Series

Primary Arithmetic. Reading, writing and understanding numbers to 100; adding and subtracting with sums to 20; multiplication and division to 10s; and signs and vocabulary needed for this level of arithmetic.

Intellectual Arithmetic. Reading, writing and understanding of higher whole numbers, fractions, and mixed numbers; addition, subtraction, multiplication and division of higher numbers; computation of simple fractions; beginning ratio and percentage; and signs and vocabulary needed for all these operations.

Practical Arithmetic. Roman numbers; carrying in addition and borrowing in subtraction; measurement and compound numbers; factors; decimals and percentage; ratio and proportion; powers and roots; beginning geometry; advanced vocabulary.

Higher Arithmetic. Philosophical understandings; principles and properties of numbers; advanced study of common and decimal fractions, measurements, ratio, proportion, percentage, powers, and roots; series; business math; geometry.

Test Examples. A supply of problems for making tests to accompany study in *Practical Arithmetic* and *Higher Arithmetic*.

Key to Ray's Primary, Intellectual and Practical Arithmetics. Answers to problems in the three lower books.

Key to Ray's Higher Arithmetic. Answers to problems in the higher book.

Parent-Teacher Guide. Gives unit by unit helps for teaching; suggests grade levels for each book; provides progress chart samples for each grade and tests for each unit.

McGuffey's Reading Series

Primer. Begins with the alphabet, moves to simple one-syllable words such as *cat* and *fox*, then on to more difficult one-syllable words such as *horse* and *spring*.

Pictorial Primer. Begins with the alphabet. First lessons have simple three- and four-word sentences with no paragraphing. The lessons progress to longer sentences and ordinary paragraphing.

First Reader. Follows the Primers at second or third grade. Words usually have the main alphabetic sounds, and few have silent letters. Helps to build fluency in using the phonics principles learned at primer levels. Stories of children who want to please God and love to learn.

Second Reader. Begins with fairly easy one- and two-syllable words and progresses to more difficult words. Reading selections on a variety of topics. Can be used to about fifth grade.

Third Reader. For students who have mastered basic reading skills and are fluent in easy reading. This Reader develops more advanced vocabulary and thinking skills. Can be used for two or three years, beginning at about sixth grade.

Fourth Reader. Develops advanced comprehension, requiring students to understand a variety of viewpoints and think about abstract ideas. Readings on themes such as life values, truth, religion, and freedom. Can be used at high school level.

Progressive Speller. Can be used at all grade levels. Begins with phonics rules. Covers spelling difficulty from one-syllable words to very complex words.

Parent-Teacher Guide. Explains how to teach reading and has specific ideas for using the lessons in this series of Readers. Also gives guidelines for helping children grow in all the language arts—spelling, writing, speaking, penmanship, grammar.

Materials for Teaching Phonics and Spelling

Phonics Made Plain by Michael S. Brunner. Flashcards for teaching sounds of the letters and letter combinations which make up the code for reading and writing. Wall chart for classroom reference and teaching of the sounds. Instructions for use.

The ABC's and all Their Tricks by Margaret M. Bishop. A comprehensive book of phonics knowledge for teachers. Useful for study, reference, solving students' spelling problems, and answering any phonics question you can think of. Tells how to teach children of all ages who have difficulty with phonics.

A Measuring Scale for Ability in Spelling by Leonard P. Ayres. The classic book from 1915, describes the research which identified one thousand most commonly used words and arranged them into difficulty groups to use for testing children's spelling level and for teaching spelling. Includes complete directions for testing.

Mrs. Silver's Phonics Workbook by Claudine Silver. A lesson on each single letter and on some digraphs such as *ck*, *sh*, and *th*. Ideas enough to spend several days on each lesson and learn it well. Each sound is correlated with a Bible verse and with science, art, and other school subjects. Can be used at kindergarten or first grade level. With the McGuffey series, it should be used before the Primers. Available in both a pupil edition and a teacher edition.